THE SAVE STUDY

RELATIONSHIPS BETWEEN THE ENVIRONMENT AND BASIC NEEDS SATISFACTION IN THE SAVE CATCHMENT, ZIMBABWE

Edited by

B. M. Campbell, R. F. Du Toit and C. A. M. Attwell

Supplement to *Zambezia*, 1988
The Journal of the University of Zimbabwe
Editor: R. S. Roberts

UNIVERSITY OF ZIMBABWE
1989

Published by

University of Zimbabwe Publications
P.O. Box MP 45
Mount Pleasant
Harare
Zimbabwe

ISBN 0-908307-08-X

The authors acknowledge the support of the United Nations Environment Programme in conducting the research for this study.

Typeset by University of Zimbabwe Publications
Printed through Print Brokers (Pvt) Ltd., Harare

CONTENTS

LIST OF CONTRIBUTORS

C. A. M. Attwell	Department of Biological Sciences, University of Zimbabwe.
B. M. Campbell	Department of Biological Sciences, University of Zimbabwe.
J. M. Clarke	Forestry Research Institute, Forestry Commission, Harare.
J. A. Codd	Department of Civil Engineering, Napier College, Edinburgh.
R. F. Du Toit	International Union for Conservation of Nature and Natural Resources, Harare.
B. H. Kinsey	Centre for Applied Social Sciences, University of Zimbabwe.
R. H. Loewenson	Department of Community Medicine, University of Zimbabwe.
D. T. Pankhurst	School of Social Studies, University of Liverpool.
I. Reh	Department of Research and Specialist Services, Ministry of Agriculture, Harare.
M. J. Swift	Department of Biological Sciences, University of Zimbabwe.
S. A. M. Van Oosterhout	Department of Biological Sciences, University of Zimbabwe.

This work represents a group effort with the final written document being compiled by the three editors on the basis of contributions from group members. The compilers and contributors for each chapter are identified; the two analytical chapters (5 and 7) were compiled from a workshop document. In a document of this length covering such a wide range of subjects and written by ten authors, there are bound to be differences in opinion; the views and conclusions expressed are not necessarily a reflection of those of all group members. The most serious differences in opinion have been recorded in footnotes by B. M. Campbell.

B. M. Campbell was the project leader and was responsible for the data analysis; B. H. Kinsey finalized questionnaire design; J. M. Clarke and S. A. M. Van Oosterhout organized the major field expedition, trained and supervised enumerators, and together with R. F. Du Toit and D. T. Pankhurst were responsible for the post-coding of the questionnaires. R. H. Loewenson's major contribution was in the early stages of the project (to questionnaire design and to the methodology concerning health and nutrition).

ACKNOWLEDGEMENTS

This book is the culmination of an interdisciplinary project sponsored, for the most part, by the United Nations Environment Programme. Additional sponsorship came from the Research Board of the University of Zimbabwe. The British Council funded the participation of J. A. Codd.

Several people, who do not appear as authors, participated in crucial aspects of the work. We especially thank Mr U. Dabholkar who, as representative of the United Nations Environment Programme, provided administrative support and criticism of draft manuscripts. Prof. M. J. Swift, Chairman of the Department of Biological Sciences, was involved with the project from implementation to completion, and we thank him for his valuable input. Dr S. Cocchi (Technosynesis, Agricultural and Rural Development Authority) and Ms K. Truscott (Agritex Socio-economic Unit) shared many experiences and ideas during the course of their work in the Save Communal Area. For critical comment on a draft manuscript we thank Prof. M. W. Murphree, Prof. M. J. Swift and three anonymous reviewers. Miss Sioux Harvey typed the original document.

GLOSSARY: LATIN–SHONA PLANT NAMES

Acacia nilotica	muunga
Acacia nigrescens	mukuu
Azanza garckeana	mutohwe
Brachystegia boehmii	mupfuti
Brachystegia glaucescens	muunze
Brachystegia spiciformis	musasa
Colophospermum mopane	masaru/mopane
Combretum apiculatum	mugodo
Dichrostachys cinerea	mupangara
Diospyros mespiliformis	mushuma
Julbernardia globiflora	munhondo
Kirkia acuminata	mubvumira
Strychnos cocculoides	mutamba

1 THE SAVE STUDY: BACKGROUND AND METHODOLOGY

B. M. Campbell, J. Clarke, R. F. Du Toit,
B. H. Kinsey, D. T. Pankhurst*

1.1 Introduction

Considerable international debate has taken place on the relationship between economic development and its environmental implications (Sandbach, 1980). Historically, developed countries that have single-mindedly pursued economic growth at any price have often discovered that such growth, if it materialized at all, was generally accompanied by direct and adverse effects upon the environment. This led to a polarized view of 'growth' and 'conservation', a view often aggravated by the different time perspectives adopted by 'growth-promoters' and 'conservationists'. Resource economics has tended to stress the maximization of short-term benefits of resource use, whereas long-term, less-quantifiable social and ecological dimensions of resource utilization have received little consideration in this discipline. In contrast, environmental sciences are often concerned with the long-term effects of factors that may show no immediate effect (Dasmann *et al.*, 1974). Resolution of this conflict has subsequently been proposed in the World Conservation Strategy, where conservation is defined as 'the management of human use of the biosphere so that it may yield the greatest sustainable benefit to present generations, while maintaining its potential to meet the needs and aspirations of future generations' (International Union for Conservation of Nature and Natural Resources, 1980).

In less developed countries the situation is complicated by the fact that underdevelopment is also a major cause of environmental degradation (Tolba, 1977). Social and economic development may contribute to the alleviation of poverty, which in turn may result in decreased environmental impact. In other cases, however, economic growth has not necessarily eliminated poverty, and relative wealth differentials have instead simply been exacerbated. Recognition of this phenomenon led to

* Compiled by C. A. M. Attwell

the basic needs approach to the development process (International Labour Office, 1976).

1.2 The basic needs approach

There are two qualitatively different approaches to the problem of defining basic needs: a) the basic material needs approach, and b) the basic human needs approach. The former stresses those needs necessary for physical reproduction, while the latter places more emphasis on those features which make life worth living in different cultures. As this latter approach invokes subjective concepts and value judgements for specific cultures, universally valid definitions are impossible. In comparison, material needs are quantifiable, and the basic material needs approach has consequently formed the mainstream interpretation of the basic needs approach. In this study it was considered that only the material needs approach would provide the objective data required to relate the needs satisfaction to the environment. This study focused specifically on two of the 'core basic needs' (Burki, 1980): health and food.

The World Health Organization defines health as a state of complete physical, mental and social well-being, and not merely in terms of absence of disease or infirmity (World Health Organization and Unicef, 1978). In this view, the need for health can be met through the satisfaction of all other needs; the level of satisfaction of health needs will thus be a useful indicator of the degree to which other basic needs are being satisfied. In addition, this study briefly considers other basic needs. Education, transport and marketing systems, even if not regarded as basic needs in their own right, must certainly be regarded as means to facilitate access to other basic needs.

1.3 Conservation and development perspectives in Zimbabwe

1.3.1 *Historical perspectives*

Riddell (1978) has reviewed the social and political factors in the apportionment of land by previous governments. Historically, land in Zimbabwe (then Rhodesia) was divided along racial lines by the Land Apportionment Act of 1930. Approximately equal areas of land were designated for White and Black farming although the Black rural population made up over 80 per cent of the total population. The provision of infrastructure and support services were heavily biased in favour of White farmers, as were marketing and credit arrangements. A strong motivating force for the setting up of the communal areas related to the supply of cheap labour for White capital investment, and hence many measures prevented effective Black competition in agriculture. (For detailed analyses of the history of the communal areas see United Nations, 1980; Phimister, 1977; Roder, 1964; Arrighi, 1970). One result of the Land Apportionment Act was that the majority of the rural population was concentrated in areas of reduced agricultural potential, often lacking essential services. Traditional agriculture had relied on extensive tracts of land on which shifting agriculture could be practised. The restricted land allocation, together with increasing population pressures, rapidly foreclosed this option.

As environmental degradation became apparent, pre Independence governments devoted considerable attention to environmental concerns, but this was often in isolation from the political, social and economic factors which themselves had contributed to the environmental problems. This period was thus characterized by a relative absence of mutually supportive measures for environmental conservation and poverty alleviation. Indeed, environmental protection measures — sometimes applied coercively — came to symbolize in the minds of the rural population the regime that had appropriated the best land and other natural resources in the first place.

The Native Land Husbandry Act of 1951 (first implemented in 1955) had as its stated objective 'the efficient use of land for agricultural purposes and for conserving natural resources and providing good husbandry'. However, it showed considerable disregard for tribal customs regulating land-use and transfer (Yudelman, 1964). While the Act aimed to intensify production, by making agriculture sedentary rather than shifting, it generated large numbers of landless who were not gainfully absorbed into the towns (Cliffe, 1981). Despite such problems, the Land Husbandry Act contained many sound environmental protection measures, to the extent that some aspects were subsequently incorporated into post-Independence environmental strategies.

1.3.2 *Current approaches*

The approach of the present government appears to be more holistic, and objectives such as the improvement of agricultural production, the generation of employment, and the improvement of education and health services have been explicitly linked in development proposals to programmes aimed at protecting the environment. Not only do major policy statements issued by the Government of Zimbabwe indicate a commitment to environmentally sound development, but governmental support has recently been given to the development of a National Conservation Strategy for Zimbabwe. Policy statements recognize that Zimbabwe's national development task is primarily concerned with the satisfaction of basic needs of the rural poor. 'Strategies are directed primarily to the amelioration of the quality of life for the rural population who are the majority and who have suffered the greatest discrimination and endured the lowest standard of income and service' (Zimbabwe, 1981a). In the sectoral analyses for the *Transitional National Development Plan, 1982/3–1984/5* (Zimbabwe, 1982b and 1983b) a basic needs approach is explicit; the statements 'the provision of adequate and decent housing and housing conditions is a basic need' and 'government recognizes that access to adequate health facilities is a basic right' are but two examples of the development philosophy underlying the Plan.

Environmental concerns are specifically included in two of the fifteen primary objectives spelled out by Government (Zimbabwe, 1981b):

(a) to develop and restructure the economy in ways which will promote rural development, desired changes in patterns of consumption, technology, exports, etc., and in ways consistent with the most desirable use and conservation of our natural resources and the environment (p. 2); and

(b) to conserve our natural resources so that production is sustained, to replace renewable resources used, and to exploit our natural resources, especially our non-renewable ones, at rates consistent with the needs of present and future generations of Zimbabweans (p. 3).

This emphasis on basic needs satisfaction, as well as on the sustainablility of resource utilization, highlights the need to view these processes as essentially interacting. The literature on environmental changes in African savanna and woodlands is extensive (see, for example, the reviews by Werger, 1978, and Huntley and Walker, 1983); however, much of this research lies outside any socio-economic context. Similarly, the information available on the satisfaction of basic needs generally lacks an environmental context. Until recent programmes such as Man and the Biosphere (Unesco, 1972 and 1975) and the United Nations Environment Programme's focus on basic needs and the environment (United Nations Environment Programme, 1982; McHale and McHale, 1978; Tolba, 1977), there has been little emphasis on basic needs–environment linkages, apart from anthropologically based studies (see Scudder, 1973 and 1980, for African examples), and generalized statements in geographical texts. This study attempts to focus on both environmental processes and on basic needs satisfaction, and to analyse the linkages between the environment and needs satisfaction.

1.4 The Save catchment study

1.4.1 *Objectives*

In the light of the above, the following objectives were arrived at:

(a) to provide detailed correlative analyses of the linkages between basic needs satisfaction and patterns of land-use, with emphasis on ecosystem perturbation and on the changes in the availability of natural resources;

(b) to collect data on the nature of peoples' perceptions of human–environment interactions, since environmental changes are in part a function of the manner in which the environment is perceived; and

(c) to specify the economic, social and environmental measures required to promote protection and improvement of the environment in synchrony with the satisfaction of basic needs. A subsidiary objective within this framework was to arrive at estimates of sustainable population densities.

1.4.2 *Choice of the Save catchment*

In addressing issues related to basic needs satisfaction and environmental change in Zimbabwe, agriculture must form the focal point. Not only do 70 per cent of Zimbabweans live off the land, but the agricultural sector makes a key contribution to the economy. In 1981, agriculture's share in the gross domestic product was nearly 18 per cent, while — with nearly 30 per cent of total formal sector employment — agriculture was the largest single employer in the country (Zimbabwe, 1982b and 1983b). This sector is an important source of industrial raw materials as well, and in 1981 it accounted for about 35 per cent of merchandise exports.

Justification for basing the study in the Save (formerly Sabi) catchment in south-eastern Zimbabwe is threefold. Firstly, the area is of national importance in Zimbabwe's agricultural development. The Save River and its tributaries are utilized for major irrigation schemes that have been established in Zimbabwe (Williams, 1981). On the Save River, these include the Chisumbanje and Middle Save schemes, established in the mid-1960s (Fig. 1.1). The importance of the Save and associated rivers to the development of the country can be gauged from an annexure to the *Integrated Plan for Rural Development* (Rhodesia, 1979), which includes statements such as 'it has long been realized that the south-eastern lowveld has by far the best potential for extensive irrigation development' (p. 17). Major developments proposed for the Save River include two large dams, Chitowe and Condo (Fig. 1.1), and an expansion of the Chisumbanje irrigation scheme, from its present area of 24 km^2 to 400 km^2. If developed to its maximum, Chisumbanje will rank among Africa's largest irrigated settlement schemes (Atkins Land and Water Management, 1983). The proposed reservoirs will each have storage capacities of around 3 000 million m^3 and will flood an area of 194 km^2 in the case of Condo and 320 km^2 in the case of Chitowe.

Secondly, the Save catchment typifies many of Zimbabwe's socio-economic and environmental problems. Most of the area falls within agro-ecological regions characterized by unreliable rainfall and sandy granite-derived soils of poor quality (see Section 1.5.4). Human populations exceed the carrying capacity (Whitsun Foundation, 1980), and environmental degradation is now widespread (Whitlow, 1980). Half of Zimbabwe's population lives in the south-east basin formed by the Save and Lundi rivers, and their future is severely threatened by soil loss, desertification and siltation. In addition to these environmental factors, poor agricultural performance in this region can be attributed to lack of credit facilities and input supply, inadequate marketing services, insufficient extension staff and relatively undeveloped infrastructure.

Thirdly, choice of the Save catchment as a study area was based on the availability of three land-tenure systems (Fig. 1.2):

Figure 1.1: THE STUDY AREA

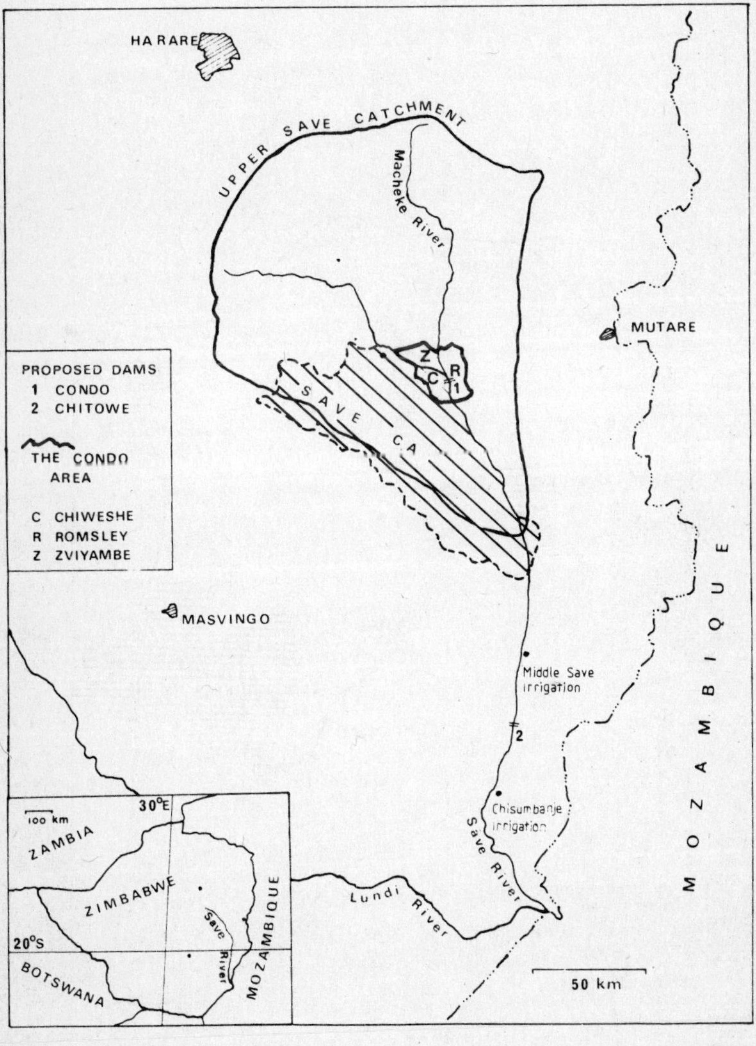

The Condo area consists of the southern half of Zviyambe SSCFA, the western half of Romsley Resettlement Scheme, and the north-eastern half of Chiweshe Ward of Save CA.

Figure 1.2: LAND TENURE SYSTEMS OF THE UPPER SAVE CATCHMENT

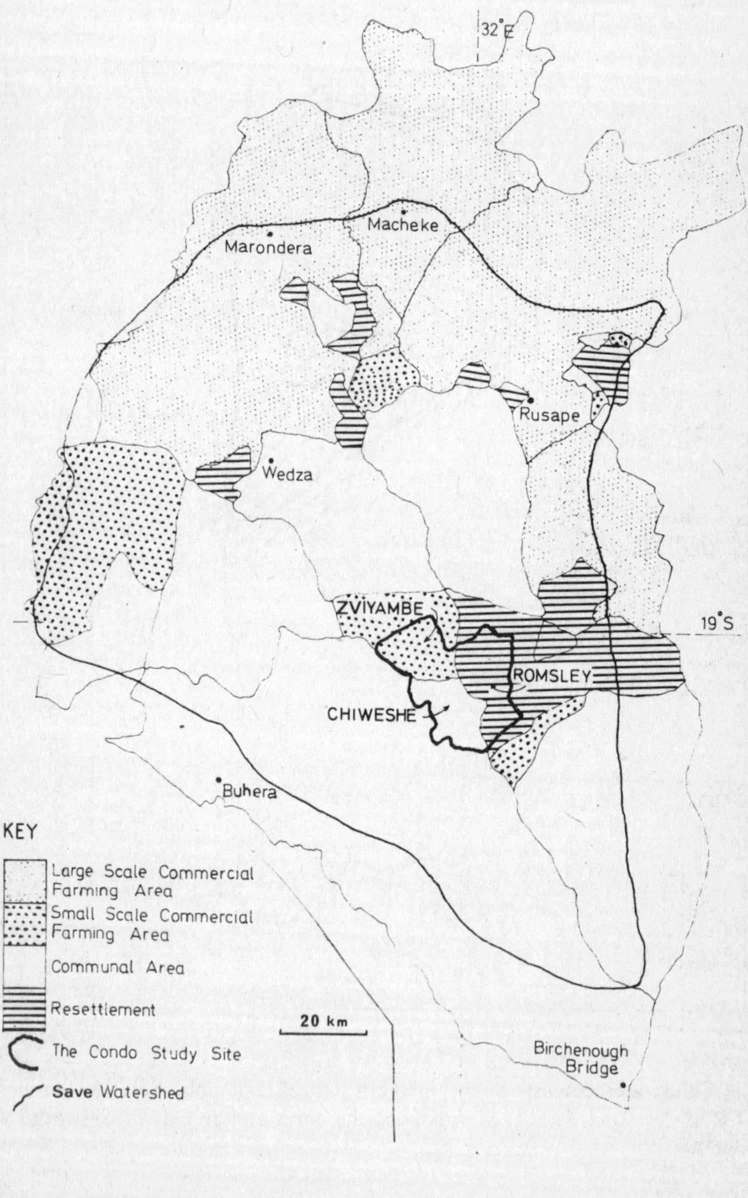

(a) Communal Areas (CAs) formerly the Tribal Trust Lands. These areas were reserved for Black farmers, and land was held under communal tenure. Much of the catchment consists of such Communal Areas.

(b) Small-Scale Commercial Farming Areas (SSCFAs) formerly African Purchase Areas, where small- and medium-scale free-holding constitutes the tenure system. This was the only land where Blacks were able to purchase and own land individually, the purpose of which was to encourage the development of a limited number of Black commercial farmers (Ndlela, 1981). About 15 per cent of the former African Area in the upper Save is devoted to SSCFAs with 85 per cent being CA — a pattern repeated throughout Zimbabwe.

(c) Resettlement schemes, initiated in 1980, where land was acquired by the state in the former European Areas (now the large-scale commercial farming areas) on a willing buyer–willing seller basis.

It was assumed that each tenure system would display a specific set of environmental and socio-economic conditions, and that such between-site variation might clarify the nature of the linkages between basic needs satisfaction and the environment. However, this study concentrates on a CA, as this area demonstrates the most severe environmental and socio-economic problems. Data from an SSCFA are included for comparative purposes. The data collected from the resettlement areas are not considered in any detail, as the very recent establishment of such systems had resulted in too great a state of flux for valid comparisons to be made.

1.4.3 *Methods*

Much of the agricultural and socio-economic data were derived from a questionnaire survey of just over 150 households, administered during 1983 in a 300 km^2 area (which included a total of about 650 households in a CA and about 200 households/farms in a SSCFA). Here, a household unit is defined as including those members normally living and eating together on the same home-site. This includes 'resident' household members who are more or less permanently in the area, as well as other members who may be working or residing away part of the time. This latter inclusion constitutes the 'self-defined' household (Truscott and Pambirei, 1983), as indicated by the household members themselves (household composition data can be found in Table 6.1). Although sample sizes are relatively small, the socio-economic data tend to agree very closely with those obtained in a census of all households in the same study area (Truscott and Pambirei, 1983).

The questionnaire examined farming operations, services, facilities, nutrition, health, social organizations and aspects of demography. Statistical analysis of the data was performed via the SPSS computer program (Nie *et al.*, 1979). Chi-squared analyses were performed on discontinuous data, and ANOVA on continuous data. In the interests of brevity, the statistical results are highly summarized, with the following indicated: the test, the degrees of freedom (d.f.), and the significance level. Classes were combined or deleted when expected frequencies were too low to apply the chi-squared test.

Interpretation of the questionnaire results was aided by information generated from less formal discussions with individuals and groups. The discussions were held with two women's groups and two farmers' groups, each comprising about 20–30 persons; the farmers' groups were all male.

Features of vegetation, geology and the extent of arable land were mapped from aerial photographs. Field survey using the Braun-Blanquet technique, in conjunction with numerical classification techniques, were used to derive a classification of plant communities (Campbell and Du Toit, in press).

Methods of demographic analysis require more specific comment, and are discussed separately in Chapter 6.

1.5 Study area

1.5.1 *Location and tenure systems*

The study area in the upper Save catchment centred on the area of the proposed Condo dam and is here referred to as the Condo area (Fig. 1.1). The Condo area covers some 55 000 ha. It consists of portions of three tenure systems: Chiweshe Ward of Save CA, Zviyambe SSCFA, and Romsley Resettlement Area (Fig. 1.2). Although data were collected in Romsley they have not in general been presented for the reasons outlined above (Section 1.4.2).

Major attention in the study is devoted to Chiweshe Ward, the northern portion of Save CA in Buhera District. Chiweshe is bounded to the west by the Mavangire Hills, to the south by the Mwerahari River, and to the north and east by the Save River (Fig. 4.2).

Some quantitative parameters of the Chiweshe study site are given in Table 1.1. In Chiweshe, the village boundaries are not entirely spatially defined, their definition being based on traditional clan relationships. Huts and fields are scattered throughout a village area with little tendency towards nucleation of huts, though there are concentrations of huts along the major roads. Fields are farmed individually but the grazing area is communal. The average area of land used for crop production by each household is 3,6 ha (Truscott and Pambirei, 1983). Although Chiweshe is part of a long-established CA, there is considerable dynamism with respect to settlement patterns, as many households have been on their present home-site for less than 10 years.

Zviyambe SSCFA (Fig. 2.1) was established in the 1950s, and most households have therefore been in the area for 25–30 years. Unlike Chiweshe, each landholding is fenced, often with internal fencing of paddocks. There is a single homestead complex on each farm. Study definitions pertaining to Zviyambe are included in Table 1.1. Data on household composition are presented in Table 6.1.

Table 1.1

GEOGRAPHICAL CHARACTERISTICS OF THE STUDY SITES IN THE CONDO AREA

	Chiweshe Ward of Save CA	*Zviyambe SSCFA*
Total area (ha)	32 700	37 780
Total number of socio-agricultural units	36 villages	570 farms
Area considered for major analysis (ha)*	16 500	17 000
Socio-agricultural units addressed in analysis	22 villages; 650 households	200 farms/ households

* The area for major analysis is that closest to the Save River.

1.5.2 *Geology and soils*
Geologically, much of the Condo area is formed of gneissic granite. This gives rise to fersiallitic soils, consisting mostly of coarse-grained sands overlying relatively unweathered granite. Such sands are inherently poor, both in fertility and in available water-capacity. In addition, dolerite intrusions give rise to reddish fersiallitic soils, which have high inherent fertility together with favourable moisture-retention properties (Thompson and Purves, 1978).

1.5.3 *Vegetation*
In its unmodified form, vegetation of the Condo area consisted principally of 'miombo' woodland on gneiss-derived soils. (Campbell and Du Toit, in press). This community has the deciduous trees *Julbernadia globiflora* and *Brachystegia boehmii* as co-dominants, which constituted about 60 per cent of the Condo area. About 20 per cent of the Condo is characterized by land which supported, and to a large extent still supports, a *Colophospermum mopane* woodland community. This community is usually restricted to drainage lines where there is an accumulation of sodium-rich bases and clays translocated from up-slope soils. Such areas are prone to erosion.

Other climax communities are represented in the Condo area, but of this remaining group no single community occupied more than 5 per cent of the total area. These communities are: *Brachystegia glaucescens* community; *Kirkia acuminata* community; *Combretum apiculatum – Acacia nigrescens* community; and an evergreen riverine community of varying species composition.

1.5.4 *Agro-ecology*
Agricultural potential in Zimbabwe can be broadly categorized according to five natural regions, based on the amount and variability of rainfall (Vincent and Thomas, 1960); local topography and soils are regarded as secondary factors that condition the land-use system recommended for a natural region. In the Save catchment, there is a trend of decreasing agricultural potential towards the south, corresponding to decreasing annual rainfall of increasing variability. The Condo area falls mostly within Region IV; Region III is confined to the north-east corner. The agricultural potential of these regions may be summarized as follows:

(a) Region III: Rainfall moderate (650–800 mm per annum) but subject to fairly severe mid-season droughts. It is marginal for intensive cropping and is better suited to livestock production.

(b) Region IV: Rainfall low (450–650 mm per annum), with severe mid-season droughts, therefore only suitable for livestock production and drought resistant fodder crops.

As rainfall is the major agro-ecological determinant, it should be stressed that the study period coincided in part with the drought of 1982/3. This season was one of the worst on record for Zimbabwe, with overall rainfall about 270 mm below normal (Torrance, 1983). Hence, the intensity of specific linkages between needs satisfaction and the environment may have been heightened over the study period.

Agricultural land, once assigned to any specific natural region, may be further classified according to the capability of the land for permanent production. In Zimbabwe, land is classified into one of eight land capability classes, whose numerical code increases as the choice of land-use becomes more limited (Ivy, 1981). For most practical purposes, the original climax vegetation communities can be equated with a specific land capability class, as shown in Table 1.2, which also indicates the relative extent of the land capability classes.

Table 1.2

THE ORIGINAL PLANT COMMUNITIES
AND THE AGRICULTURAL CAPABILITY CLASSES
IN THE STUDY AREA

Original plant community	*Capability class*	*Area (%)* Chiweshe	*Area (%)* Zviyambe
C. apiculatum – A. nigrescens (plains)	II	12	0
J. globiflora	III	50	73
C. mopane	(IV)/V	20	21
B. glaucescens	VII/VIII	3	4
K. acuminata	VII/VIII	1	3
C. apiculatum – A. nigrescens (hills)	VII/VIII	13	0

For the Condo area as a whole, a maximum of about 65 per cent of the land is classed as arable, falling into capability classes II and III. Class II is characterized by moderate limitations to cropping, and Class III is characterized by severe limitations. In land falling within Class III (the most extensive class in the Condo area), cropping should only be done with intensive protection measures. At least 2 per cent of the total land area is devoted to huts and roads; other non-arable features which were not mapped include small rock outcrops and shallow soils. If all these are taken into account, then the maximum land available as arable is probably closer to 50 per cent in Chiweshe and 60 per cent in Zviyambe.

Drainage lines cover about 20 per cent of the Condo area and fall into capability Class V. As such, they are mostly not arable, although small portions fall within capability Class IV. This latter class includes land subject to very severe permanent limitations for cultivation, and cropping would normally be considered only under intensive protection regimes. About 10 per cent of the Condo area is hilly country in either Class VII or Class VIII, both suited only for rough grazing and wildlife.

2 RESOURCE USE AND AGRICULTURAL PRODUCTION

I. Reh, S. A. M. Van Oosterhout,
B. M. Campbell and R. F. Du Toit *

2.1 Land-use patterns

The land-use system in Chiweshe under past conditions of low population densities was traditionally that of shifting agriculture. The vegetation over a small patch of land was cut down and burnt. Soil fertility was temporarily increased when the cultivators spread the ashes (rich in phosphate, potash, liming materials and absorptive charcoal) over the soil. The heat of the fires sterilized the soil and increased the availability of nitrogen and sulphur (Grant, 1981). After use of a patch for 2–3 years, fertility was depleted and the patch was allowed to revert to a bush fallow for the next 15 or more years. The presence of leguminous shrubs such as *Crotalaria* indicated that soil fertility had been replenished to the extent that cultivation was again worth while. The total area of cropland and fallow required per family was 150–200 ha, and provided this land was available the system was self-sustaining and did not degrade the soil (Grant, 1981).

With population increase, this system of cultivating small patches within a mosaic of regenerating woodland was no longer feasible and the pattern of land-use has changed considerably. Interpretation of aerial photographs indicates the extent of land cleared for cultivation (i.e. land with a cultivation history) is over 50 per cent in Chiweshe and about 40 per cent in Zviyambe (Fig. 2.1). However, responses from farmers who were approached in the questionnaire survey indicate that less than 12 per cent of the land area in both Chiweshe and Zviyambe was cultivated in 1982/3. These land areas, together with small patches which the farmers said they had fallow, make up the total cropping area cited by farmers: only about 15 per cent of the total area within both Chiweshe and Zviyambe. Clearly, there are large areas of land which have been cleared

* Compiled by R. F. Du Toit.

Figure 2.1: LAND USED FOR CULTIVATION

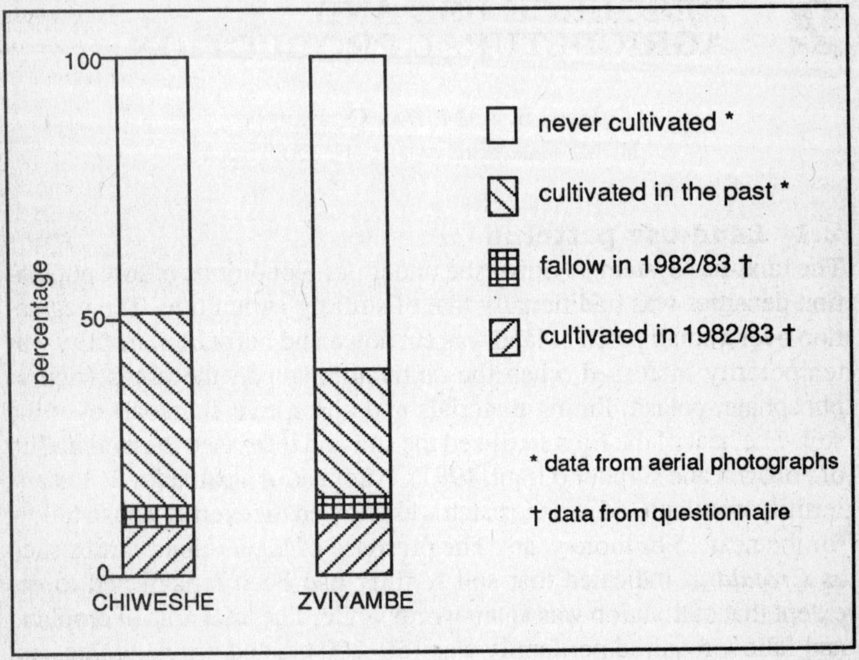

at some time in the past but which the farmers no longer regard as being fallow land with some potential for future cropping.

A further insight into the degree of land-use is provided through a comparison of the areas of land which can be classified as arable according to soils and slopes (see Section 1.5.4), and those which have actually been cleared for cultivation. In Zviyambe, about 60 per cent of the land is arable, and therefore present utilization is well within the potential. In Chiweshe, however, a maximum of about 50 per cent of the land is arable whereas the present extent of cleared land is already at least 50 per cent; this indicates the possibility of some use, either presently or in the past, of land that is marginal for cropping. It was found that this has in fact occurred, usually with land that was formerly covered by the *Colophospermum mopane* community and by 'vlei' vegetation (treeless grasslands of drainage lines). Considering the Zimbabwean Natural Regions classification based on rainfall (see Section 1.5.4), there is again an indication of over-use of marginal land. The large cleared areas in both Zviyambe and Chiweshe show an intensity of cropping that is not in keeping with the environmental limitations characteristic of Natural Region IV; extensive livestock production on rangeland is the most ecologically appropriate agricultural activity in such an area.

Environmental implications of the emphasis on crop production are outlined in Chapter 3. Returning to the evolution of the current land-use pattern in Chiweshe and much of the rest of the upper Save catchment, it was found through the study of aerial photographs that the area of fields has increased considerably in Zviyambe since the mid-1970s whereas the increase in Chiweshe has been slower (Fig. 2.2). This is no doubt because the maximum possible extent of cultivation has been reached in

Figure 2.2. CHANGE IN CULTIVATED AREA

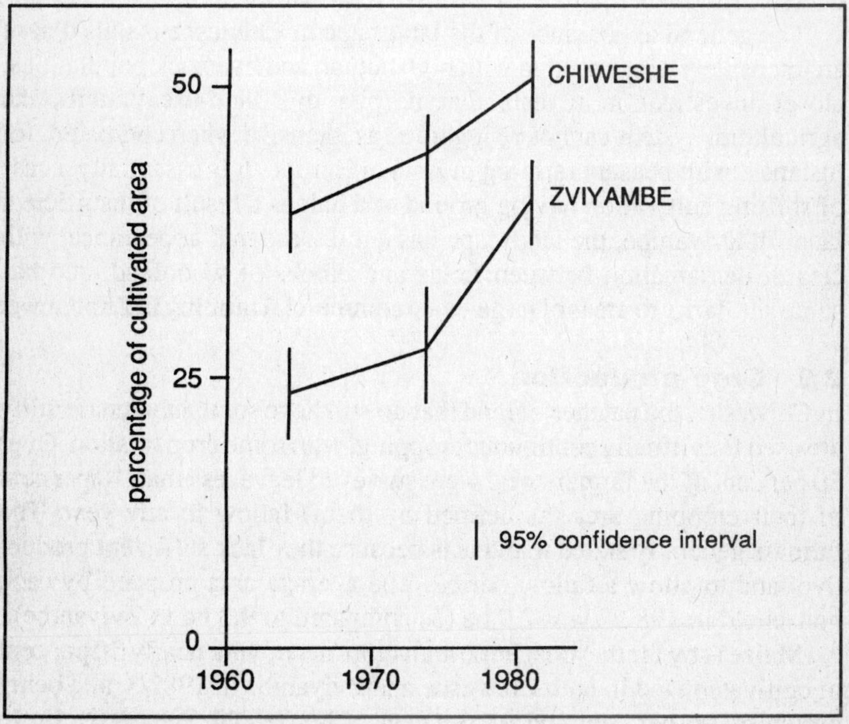

Data from aerial photographs.

the CA. Overall, therefore, there has been a trend towards more intense use of land for cropping — a trend that is to be expected as population density increases; cropping limits have been reached some time ago, and there has been some abandonment of overutilized land.

The areas of fallow, abandoned or uncleared land which are not used for hut sites, roads and other features of human settlement are given over to grazing for cattle and small livestock. In Chiweshe, patches of rangeland in areas of gentle topography, particularly near villages, have had their vegetation cut in a somewhat haphazard fashion by wood collectors and are criss-crossed with cattle paths. These degraded areas have a very different appearance on aerial photographs and satellite images from the hilly areas, rising above the settled plains, which still have a good tree cover. Cultivation often extends into drainage lines, so the watercourses have only narrow, discontinuous strips of trees such as *Colophospermum mopane* to stabilize their banks.

The general appearance of the landscape in Chiweshe is thus that of an intensively-cropped area with high human and livestock populations; closer investigation reveals that despite this land-use pattern, the agricultural system cannot be regarded as intensive when compared, for instance, with peasant farming in Asia or Europe. It is essentially a case of shifting cultivation having ground to a halt as a result of insufficient land. In Zviyambe, the landscape has a more ordered appearance, with clearer demarcation between fields and blocks of woodland, and has some similarity to areas of large-scale commercial farming in Zimbabwe.

2.2 Crop production

In Chiweshe, the patches of land that do still have some inherent fertility are used for virtually continuous cropping, with some crop rotation. Over 50 per cent of the farmers who were surveyed leave less than 10 per cent of their cropping area (as defined by them) fallow in any year. The farmers generally stated that this is because they lack sufficient productive land to allow a fallow period. The average area cropped by each household in 1982/3 was 2,9 ha (as compared to 9,3 ha in Zviyambe).

Maize is by far the most important crop, occupying nearly 50 per cent of cultivated land in both Chiweshe and Zviyambe in 1982/3, and being grown by 90 per cent and 98 per cent of households, respectively. In general, SSCFAs have maize yields which are 3,5 times higher than those of CAs,* but in the 1982/3 season yields were exceptionally poor, with no significant differences between Chiweshe and Zviyambe. The average maize yield was 200 kg per ha, with over 70 per cent of growers having zero yields. The average yield was half that recorded in the 1981/2 season, which in turn was less than half that of the wetter season of 1980/1. Similar declines in yields occurred with other crops, as indicated in Fig. 2.3. The extent to which some of these other crops were planted in 1982/3, and the yields of these crops, are shown in Table 2.1.

Bulrush millet (mhunga) is said to be the favourite food crop in Chiweshe (Truscott and Pambirei, 1983) and is clearly a subsistence crop, as are sorghum and finger millet. Groundnuts may be grown for sale, in years of adequate rainfall, by a considerable number of Chiweshe farmers, and by most Zviyambe farmers; 10 per cent of the cultivated area of Chiweshe was taken up by this crop in 1982/3, and 25 per cent of the Zviyambe cultivated area. The greater orientation of Zviyambe farmers towards cash cropping, rather than merely subsistence production, is also reflected in the areas devoted to cotton.

Less important crops are edible beans and rice. All households in Zviyambe and 87 per cent of those in Chiweshe have vegetable gardens, within which green leafy vegetables are invariably produced. Tomatoes and onions are grown in more than 40 per cent of these gardens.

In normal years, average crop retention in Zviyambe is twice that in Chiweshe. Callear (1984) found that households in Wedza used an average of eight bags of maize per annum. After the 1982/3 season in the Condo area, households retained only an average of three bags of maize, which is certainly insufficient to feed an average of 6–8 household members.

A fairly standard five-year crop rotation in the CA is three years of maize, followed by one year of groundnuts and then a year of sorghum. The basic tillage system involves winter (dry-season) ploughing, with further ploughing or harrowing at the commencement of the rains (November–December). Animal draught is generally used for ploughing in Chiweshe, although a number of households (15 per cent in the case of maize and 30 per cent in the case of groundnuts) have no access to animal draught and have to rely on hand hoeing. A smaller percentage of

* Data from Agritex estimates for the upper Save catchment.

Figure 2.3: CROP YIELDS IN THE CONDO AREA

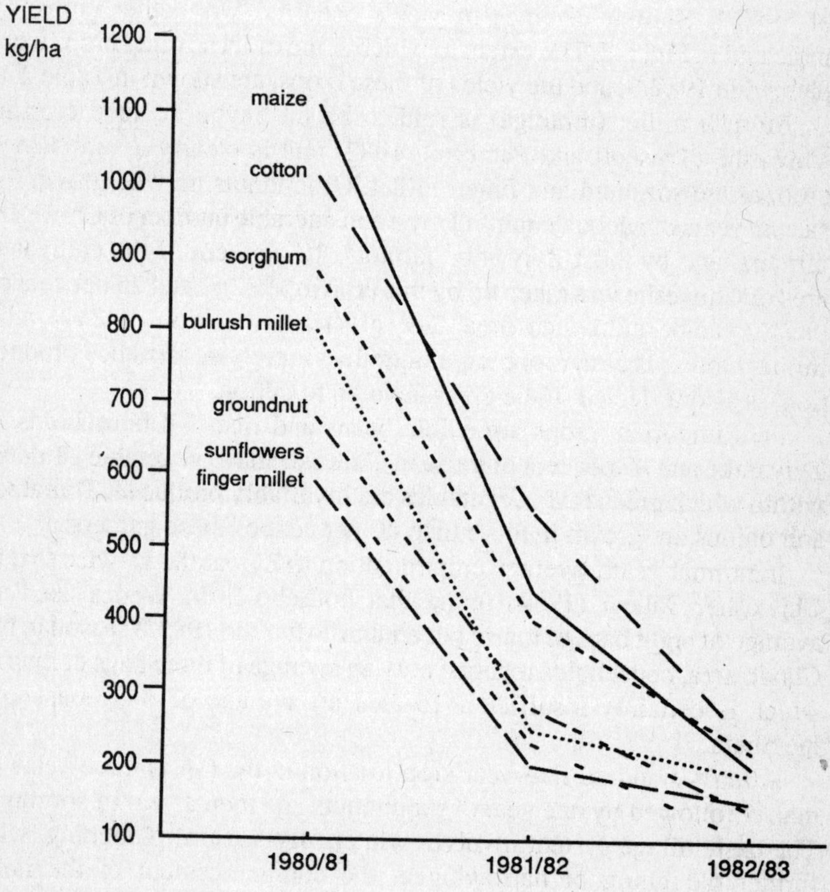

The 1981/2 data are for Chiweshe Ward (Truscott and Pambirei, 1983).

Table 2.1

PRODUCTION OF MAJOR CROPS IN CHIWESHE AND ZVIYAMBE IN 1982/3

		Maize	*Bulrush millet*	*Groundnuts*	*Sorghum*	*Cotton*
Households growing crop (%)	Chiweshe	90	45	45	36	14
	Zviyambe	98	19	93	15	47
		n.s.	**	***	**	***[a]
Mean area cultivated per household (ha)	Chiweshe	1,3	0,5	0,3	0,5	0,1
	Zviyambe	4,6	0,2	2,5	0,3	1,0
		*	*	*	n.s.	*[b]
Mean area cultivated by growers (ha)	Chiweshe	1,4	1,0	0,7	1,4	1,0
	Zviyambe	4,8	1,2	2,8	1,9	2,3
		*	n.s.	*	n.s.	n.s.[c]
Mean production per growing household (in bags; cotton in bales)	Chiweshe	4,2	2,7	2,7	4,8	1,2
	Zviyambe	8,4	1,4	6,3	1,0	1,8
		*	n.s.	*	n.s.	n.s[c]

Chiweshe (*n* = 112); Zviyambe (*n* = 53). Significant differences between areas are shown as follows:
n. s. = p > 0,5 (not significant); * = 0,01 < p < 0,05; ** = 0,001 < p < 0,01; *** = p < 0,001.
[a] (all d.f. = 1) [b] (all d.f. = 163) [c] (d.f. dependent on no. of growers)

Zviyambe households (three times smaller) rely on hand hoeing. In the case of weeding in Chiweshe, animal draught is much less used; about half the households weed the fields manually.

Various woody plants are cultivated in Chiweshe and Zviyambe to provide fruit, firewood, building material, twine, etc., to make hedges, or to be ornamental (Fig. 2.4). Whereas almost 50 per cent of Chiweshe households grow no woody plants, all Zviyambe households that were surveyed grow at least one woody species, with most growing three to five.

2.3 Ownership of livestock

The percentages of households that were found (through the questionnaire survey) to own cattle of various types are shown in Table 2.2. With those Chiweshe households which owned cattle, the average herd size is nine (range 1–40). In Zviyambe, cattle ownership is significantly higher, with surveyed households all owning some cattle (average 22; range 6–50), and 90 per cent having a minimum of two draught cattle.

Goats are owned by far more households than sheep are, but both are less prevalent than cattle (Table 2.3). Ownership of small livestock by households in Zviyambe is generally more frequent than in Chiweshe. Very few households keep pigs (for example, 2 per cent in Chiweshe).

2.4 Grazing land

The prime feature to note with regard to use of grazing land is that in CAs this is regarded as a communal asset, to be used freely by all those with livestock, unlike arable land which is apportioned amongst individual households.

In Chiweshe the possibilities for grazing include:

(a) hilly country (about 17 per cent of the total area);

(b) plains and drainage lines never previously cultivated (30 per cent); and

(c) previously cultivated areas that are not presently cultivated (35 per cent).

It appears that present grazing mostly takes place in categories (a) and (b) (Truscott and Pambirei, 1983), giving a stocking rate for cattle of about

Figure 2.4: PERCENTAGE FREQUENCY OF WOODY PLANTS CULTIVATED AT HOME-SITES

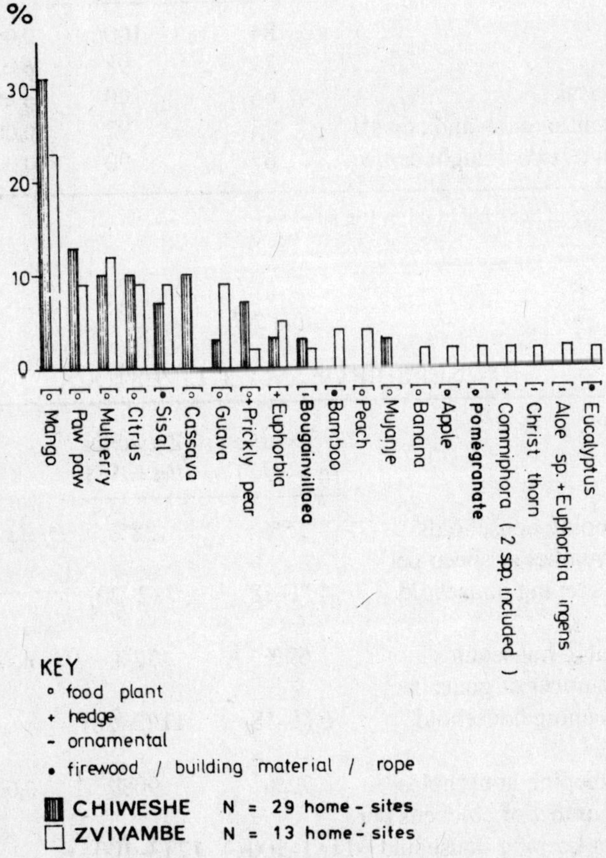

Table 2.2

OWNERSHIP OF CATTLE

	Chiweshe (n = 112)	*Zviyambe (n = 53)*	
All cattle	84	100	0,001 < p < 0,01
Cows	71	98	p < 0,001
Trained oxen	66	90	p < 0,001
Draught cattle (oxen and cows)	73	92	0,001 < p < 0,01
Minimum of two draught cattle	67	90	0,001 < p < 0,01

Percentage of households; d.f. = 1 in all cases.

Table 2.3

OWNERSHIP OF SMALL LIVESTOCK

	Chiweshe (n = 112)	*Zviyambe (n = 53)*	
Sheep-keeping households	25%	38%	n.s.
Average number of sheep per sheep-keeping household	4 (1–18)	7 (2–20)	
Goat-keeping households	69%	72%	n.s.
Average number of goats per goat-keeping household	6 (1–18)	11 (2–28)	
Chicken-keeping households	72%	90%	0,001 < p < 0,01
Average number of chickens per chicken-keeping household	11 (1–100+)	17 (3–100+)	
Chicken-keeping households			
with more than 20 birds	12%	33%	0,001 < p < 0,01
with more than 50 birds	5%	2%	n.a.

Ranges in parentheses; d.f. = 1 in all cases.

n.a. = Chi-squared analysis was not possible as frequencies were too low.

2 ha per livestock unit.* Truscott and Pambirei (1983) derive an even higher stocking rate of 0,6 ha per livestock unit (they base their calculations on a grazing area defined by farmers). If all the above categories of potential grazing land could be used, the stocking rate for the present herd would be about 3,8 ha per livestock unit. The cattle population of Chiweshe has diminished to a considerable extent from previous levels as a result of the drought (the Agritex extension worker for Chiweshe states that about half the cattle were lost during the period 1981–4), so past stocking rates were higher.

Government pasture specialists recommend that the stocking rate in the Chiweshe-type environment should be 6 ha per livestock unit (Zimbabwe, 1981). According to this criterion, Chiweshe is considerably overstocked. Assuming that maximum use of the land could be made for grazing (i.e. about 80 per cent), even the present low average cattle ownership per household of 5,4 livestock units would have to be reduced to about 3,4 livestock units per household if the recommended stocking rate is to be achieved.

Considering Zviyambe as a whole (which is not strictly valid as each farm is fenced), the recommended stocking rate is also exceeded, although not to the same extent as in Chiweshe; if all possible grazing areas (about 80 per cent of the total) are utilized by the Zviyambe herd, the stocking rate would be about 4,6 ha per livestock unit. During the drought years, the stocking rate in Zviyambe was increased since cattle from Chiweshe were herded into parts of Zviyambe.

The above analyses are based on rough grazing alone. At the time of the survey, there were no paddocks, rotational grazing schemes, improved pastures or any such development to raise the carrying capacity within Chiweshe (although two villages within Chiweshe are to be involved in a rotational grazing scheme during 1986). Paddocking does occur in Zviyambe, with areas being reserved for winter grazing.

2.5 Use of livestock

Cattle serve a multipurpose role in Zimbabwean CAs and it is not possible to appraise their management in CAs in the same way that cattle management on commercial farms is assessed.

* Livestock Unit = 500 kg animal; each adult local beast = 0,7 livestock unit.

As noted by Sandford (1982), CA farmers keep livestock primarily to meet their subsistence requirements. Provision of draught power and manure is of major importance; milk is also a valued output. Cattle may be sold to meet immediate requirements for cash, such as when remittances from family members working elsewhere are not received. Social considerations, relating to prestige, bridewealth and ritual, play a secondary role but are nonetheless important.

The multipurpose use of cattle is not readily quantifiable in monetary terms, and it is difficult to estimate the productivity of the Chiweshe cattle herd. Some indication can be obtained by looking at calving rates, mortality and off-take (Table 2.4); however, the figures are rough, and the situation is confused owing to the effects of the drought. Production performance is low, with herd losses due mainly to low calving rates and high mortality rates; the calving rates can be compared with those of other CAs (pre-drought), presented by Sandford (1982), which are generally in the range of 40–50 per cent.

Off-take is affected by the high mortality and is therefore very low, but would be low anyway since sales and slaughterings are not the prime reason for keeping cattle, unlike on the large-scale commercial farms. In Zviyambe, the herd is somewhat more productive, with higher calving rates and offtake, and apparently lower mortality. Off-take was largely accounted for by home slaughterings. The mortality patterns in both areas are similar; the highest mortality occurring with cows and calves (Table 2.5).

Measurement of the benefits of subsistence forms of livestock husbandry may be through the use of biological measures of herd productivity (Table 2.4), or through the use of economic measures on the profitability of the herding enterprise, or through an assessment of the dietary benefits gained by the herders; in fact, all such factors have to be taken into account (Behnke, 1985). With the complexity of such an exercise being much increased because of the prevailing drought situation, this survey could not investigate this important topic in any depth. The extent to which the Chiweshe herd is meeting requirements for draught and manure is discussed in Section 2.6, and its importance as a protein source for the farmers is touched upon in Section 4.3.1.

The situation with small ruminants is very similar, although meat supply for home consumption and small cash income appear to be the

Table 2.4

AVERAGE CALVING RATES, PURCHASES, MORTALITY AND OFF-TAKE FOR CATTLE IN 1982/3

	Chiweshe ($n_1 = 72$)* ($n_2 = 774$)	*Zviyambe* ($n_1 = 41$) ($n_2 = 956$)	
Calving rate (%) (no. of births related to the average annual no. of cows)	29	38	$0{,}01 < p < 0{,}05$
Purchases (%) (in relation to calculated herd size of 1981/2)	3,5	0,4	n.s.
Mortality (%) (in relation to calculated herd size of 1981/2)	21	11	n.s.
Off-take (%) (including sales and slaughterings)	3	7	$0{,}01 < p < 0{,}05$
Herd-size change over the year (%)	–17.5	–6.0	$0{,}01 < p < 0{,}05$

*Sample size n_1 equals number of households for which the full data on cattle were available; n_2 shows the number of cattle considered in the statistics.

Table 2.5

PERCENTAGE DISTRIBUTION OF CATTLE DEATHS ACCORDING TO CATTLE CATEGORIES

	Chiweshe ($n_1 = 72$)* ($n_2 = 774$)	*Zviyambe* ($n_1 = 41$) ($n_2 = 956$)
Cows	48	40
Calves	24	27
Trained oxen	12	12
Heifers	10	14
Steers	4	3
Bulls	2	4

*Sample size n_1 equals number of households for which the full data on cattle were available; n_2 shows the number of cattle considered in the statistics. There is no significant difference between areas; (n.s., d.f. = 4).

major reasons for keeping sheep and goats. Goat mortality was found to be in the order of 19 per cent in Chiweshe (surprisingly high in view of the hardiness of these animals, even under drought conditions), while off-take was about 12 per cent with three-quarters of this being home slaughterings. As with cattle, goat mortality in Zviyambe is lower (about 11 per cent) and off-take higher (21 per cent), with most of this again being due to home slaughterings. The small numbers of sheep made the derivation of such measures of productivity unreliable in their case.

2.6 Constraints on agricultural production

Apart from the somewhat adverse natural environment, aspects of human demography, economic forces, past political circumstances and inappropriate agricultural practices combine to keep the interlinked production of livestock and crops at a low level in Chiweshe. Some of the socio-economic factors found to directly affect agricultural production are outlined below.

2.6.1 *Draught shortage*

Despite the fact that the Chiweshe area is seemingly overstocked (Section 2.4), the farmers suffer from a severe shortage of draught power. If two healthy cattle are regarded as a minimum draught unit, only about 67 per cent of Chiweshe households could potentially have this power (whereas 90 per cent of the surveyed households in Zviyambe had such a minimum unit, and several had tractors — not found in Chiweshe). In fact, the majority of farmers considered that, since their cattle are in such poor condition when needed for summer ploughing at the beginning of the rains, a minimum of four cattle are required to constitute a draught unit. The shortage of draught is highlighted by the use of cows for pulling implements by over half of the Chiweshe households (which reduces their productive performance), whereas the majority of Zviyambe farmers use oxen. The problem is overcome to some extent through the sharing of available draught cattle, but this is only a partial solution as it inevitably leads to late planting (see Section 2.6.3; Callear, 1984). Many researchers have recorded a correlation between crop yields and cattle ownership (e.g. Shumba 1984a), but it is often unclear whether it is manure shortage (see Section 2.6.2) or draught shortage which is the critical constraint. Shumba (1984a) concludes that commercial crop production by CA farmers is associated with draught power ownership. Bratton (1984) regards manure shortage rather than draught shortage as the critical factor affecting crop production.

2.6.2 *Manure shortage*

About 18 per cent of Chiweshe farmers were found to use fertilizer, and 44 per cent manure, on their maize crops. About half the farmers used either manure or fertilizer. In Zviyambe, almost all farmers use either manure or fertilizer on maize. With other crops (both subsistence and cash), there is the same trend of much greater use of either artificial or natural fertilizers in Zviyambe.

Farmers in Chiweshe have stated that fertilizer is not particularly useful under the local dry conditions (Truscott and Pambirei, 1983). This is to be expected with the late planting and shallow ploughing that occurs (see below). Most of these farmers cannot risk the cash outlay for fertilizer in case of drought-induced crop failures, and so often only manure is applied to maize fields. Although moisture retention of sandy soils is improved through the addition of organic matter, Grant (1976) found that manure applied to exhausted soils (abandoned fields) is not effective in raising nutrient status without the addition of artificial fertilizers. However, if the land under cultivation was to receive all its fertilizer needs (let alone the abandoned land), the cost would be very high and the finance — either cash or credit — is not presently available. Thus the dependency on manure, at least to slow the rate of nutrient depletion, will remain unless there is some kind of aid intervention.

Experiments at a Zimbabwean research station have indicated that about 8 tonnes of manure per hectare are required annually to maintain high maize yields on granitic soils (Roder and Hopley, 1973). With a livestock unit producing about 1,3 tonnes of manure per annum, six livestock units would be required for each hectare of maize. With the average livestock holding per Chiweshe household being five livestock units, insufficient manure is produced by these for even one hectare of maize. To achieve the recommended application, at the recommended stocking rate, 36 ha of grazing would have to be set aside for each hectare of cropland (or 97 per cent of the total area). Hence, sufficient manure can never be obtained from the area of rough grazing within Chiweshe, and the only possible improvement is through the establishment of pastures or the production of fodder crops to raise livestock numbers and increase manure production (as well as to increase draught power). Quality of manure also has to be taken into account and this relates directly to the quality and quantity of fodder (M. J. Swift, pers. comm.).

2.6.3 *Late planting*
The CA farmers generally do the final ploughing and plant crops only after the first rains have fallen, by which time the cattle are thin and weak and the soil very hard. The consequent shallow ploughing leads to poor water retention; the available water at these shallow depths is rapidly depleted, and poor yields result (Shumba, 1984b).

2.6.4 *Inappropriate seed variations*
Since the 1950s, yields in the CA farming sector have almost doubled — rainfall variation excluded (Norton, 1983). This increase is mostly a consequence of research into new varieties aimed at improving commercial outputs (which have risen to the order of 4 500 kg per hectare in the case of maize in the large-scale commercial farming area of the upper Save catchment). Despite the increase in CA yields, there is still much that could be done to ensure that farmers are provided with the most appropriate seed varieties. Local traders often stock maize varieties other than short-season, drought-resistant varieties such as R200 and R201 (Gapare, 1976), or seeds arrive so late that farmers use their own. These have a very low germination rate, especially in the case of groundnuts ('certified seed' for groundnuts was not available in the CA for two years prior to the 1982/3 season). Some of the new maize varieties are not very disease/pest resistant, and are only high-yielding when treated with insecticide. It would appear that great benefit could be gained from research on traditional crops (such as millets and sorghum) which might be higher-yielding and more resistant to pests in the Chiweshe-type environment.

Other socio-economic constraints to agricultural production have been discussed elsewhere. They include equipment shortages (see Section 4.1.1), crop pricing policies (Muir, 1981), problems experienced by women farmers (see Section 5.4; Truscott and Pambirei, 1983; Truscott, 1985), insufficient agricultural extension and veterinary assistance (see Section 4.1.2), and the nutritional/health status of the farmers (see Section 4.3).

2.7 Use of woodland resources

All households interviewed in both Chiweshe and Zviyambe use indigenous wood for fuel and fencing, and most use wood for the construction of dwellings (pole-and-mud construction). The responses to a question on the ease with which firewood can be found are shown in Table 2.6.

Table 2.6

ASSESSMENT OF EASE OF ACCESS TO FIREWOOD

	Chiweshe *(n = 113)*	*Zviyambe* *(n = 58)*
Very difficult to find	9	0
Quite difficult	36	5
Quite easy	38	57
Very easy	18	38

Percentage of respondents.
There is a significant difference between areas (p < 0,001; d.f. = 2).

Thus some 40 per cent of Chiweshe respondents perceive some degree of difficulty in obtaining wood; in Zviyambe, the situation is better, with almost a third of respondents finding wood quite easy to obtain, and two-thirds finding it very easy. Group discussions with women revealed an appreciation of the fact that the supply of wood fuel is decreasing. The drought dominated the explanations proffered to explain the decline in supply. It appeared that some of those who had suffered a loss of income due to the drought were attempting to make money by selling firewood, mainly in the nearby population centre of Dorowa. This greater demand for local wood supplies seemed a more logical reason for stating that wood is becoming more difficult to obtain than another explanation based on the observation that young trees had died in the drought.

The allocation of increasing proportions of a family's labour resources to fuel gathering was established in the survey. In Chiweshe, data were collected on perceptions of the time taken and distance travelled to collect fuelwood in the past (defined as ten or more years ago) and in 1983. Respondents recalled that in the past a trip to collect firewood for the household took only twenty minutes; in 1983, the estimated time was two hours. Similarly, respondents recalled that the average distance travelled to collect firewood in the past was less than a fifth of a kilometre; now the distance travelled is reckoned to be some two kilometres further.

Another indication of progressive deforestation in Chiweshe is the range of species that are currently used. In Zviyambe, which is still relatively well-wooded, the most utilized species is *Julbernardia globiflora* followed by *Brachystegia boehmii* and *B. spiciformis*. All three species are major constituents of the climax 'miombo' woodland of the region. In Chiweshe, the most important trees are *Combretum* species and *Colophospermum mopane*. The species selection in Chiweshe reflects the vegetation that remains in this area — drainage line vegetation, regrowth on old fields, trees remaining in cultivated areas (Campbell and Du Toit, 1988).

A consequence of the diminishing availability of wood in Chiweshe is the collection of wood by the Chiweshe residents in adjacent areas: about 50 per cent of respondents were found to collect construction poles in the Romsley Resettlement Scheme, to the east, and about 30 per cent collect in Zviyambe. Another possible consequence is the shift towards the use of bricks rather than poles in the building of dwellings: 36 per cent of Chiweshe respondents said that bricks were used because of the shortage of good poles. Most respondents stated that they used bricks because they were more durable than wood. Brick walls are also required to support the asbestos roofing that is now needed in the absence of sufficient thatching grass (see Section 2.8). Shelter is further discussed in Section 4.1.4.

Over 95 per cent of people in Chiweshe and Zviyambe said that their diet was supplemented by fruits from wild trees and shrubs. The remaining few percent stated that the drought had reduced fruit production and that therefore they did not collect wild fruits. In contrast to the differences in species used for fuel in Chiweshe and Zviyambe, there are no differences in the selection of wild fruits in the two areas. The major fruit tree species are *Diopyros mespiliformis*, *Strychos cocculoides* and *Azanza garckeana*. Despite the fact that almost all wild fruits are from trees, there appears to be no effect of deforestation on fruit choice, although Chiweshe people did suggest that fruits are becoming less common. It transpired that fruit trees are selectively left in the field areas when these are cleared and are the last trees to be felled for firewood (see Campbell, 1987, for details on the use of wild fruits in Zimbabwe).

2.8 Use of grass

Grass is used for thatching by 70 per cent of the Chiweshe households, apart from being the resource on which livestock production is based. A number of households use reeds to make mats.

During the winter months of 1983 there was no measurable amount of grass on the Chiweshe rangeland, where mean values of at least 100 g per m^2 were expected. The people obtain their thatching grass from other areas, such as hills in the Romsley Resettlement Scheme, and reeds for mats have to be obtained from even more distant localities (for example, rivers near Mutare). The degradation of the grass resource is indicated by the very low frequency of past bush fires cited by Chiweshe people (whereas there were fires in nearby Romsley during the course of the study). Perceptions of respondents concerning the ease with which thatching grass can be obtained are shown in Table 2.7. The major reason stated for the difficulty in obtaining this building material was the drought, but it appears that preoccupation with the drought is masking the underlying and long-term reasons.

Table 2.7

ASSESSMENT OF EASE OF ACCESS TO THATCHING GRASS

	Chiweshe (n = 113)	*Zviyambe (n = 58)*
Very difficult to find	79	52
Quite difficult	12	24
Quite easy	7	16
Very easy	2	8

Percentage of respondents.
There is a significant difference between areas ($0,001 < p < 0,01$; d.f. = 2).

3 ENVIRONMENTAL DEGRADATION

R. F. Du Toit, B. M. Campbell

3.1 Historical information

A historical perspective on environmental change in Chiweshe and adjacent areas can be gained from the recollections of the older inhabitants of the area and from archival material (official documents and newspaper reports). Evidence of desertification cited by older respondents in the questionnaire survey included the observations that the Save River used to have big pools (inhabited by hippopotamuses) which are now dry and filled with sand, and that springs and sacred water places in Zviyambe and Chiweshe had run dry even before the recent drought. In Zviyambe, farmers complained that the vleis in which they were able to grow rice in the past are now too dry and are covered with thorn shrub.*

Native Department reports of the 1930s commented on population 'congestion' and overstocking in CAs in the northern portion of the Save catchment. In the 1940s, mention was made in such reports of shortages of thatching grass, of gully erosion, and of the fact that the source of water for many of the people was of the order of six miles from their kraals. A newspaper article (in the *Umtali Post*) of 1960 bemoaned the 'giant of erosion that has turned the Sabi into a river of sand and which would choke the vital storage dams with silt'.

In the early 1960s, official concern was such that a Save Catchment Investigation Team was appointed by the Ministry of Agriculture to assess the erosion situation in the upper sections of the catchment. The report produced by this team (Hudson *et al.*, 1964), described 'negligible' erosion in the large-scale commercial farming areas, and 'severe' erosion in the CAs falling within Natural Regions III and IV, to about 19°30′ S (south of Chiweshe). Cleghorn (1966) confirmed that grazing conditions in this region were extremely poor even in the early 1960s.

*Vleis are wetlands in headwater valley sites, and are also known as dambos in Zambia and Malawi.

3.2 Deforestation

The extent of deforestation in Chiweshe area can be investigated both from the study of aerial photographs taken in different periods and from a comparison of the woodland cover and vegetation types in this and adjacent areas. The latter aspect has been discussed, in relation to trees used for firewood, in Section 2.7. Data interpreted from aerial photographs taken in 1964 and 1981 can be compared in Fig. 3.1. A considerably greater proportion of the vegetation of the Chiweshe area has less than 25 per cent canopy area than is the case with Zviyambe, but the extent of deforestation in the CA has not altered greatly over the last 20 years. In Zviyambe, the woodland has been considerably thinned since the 1960s. This indicates that much of the land clearance in Chiweshe took place some time ago, a situation which is common to other Zimbabwean CAs (Du Toit *et al.*, 1984). The relatively slow rate of deforestation in Chiweshe over the past two decades correlates with the slow increase in cultivated area (see Section 2.1).

Figure 3.1: PAST AND PRESENT WOODY PLANT COVER

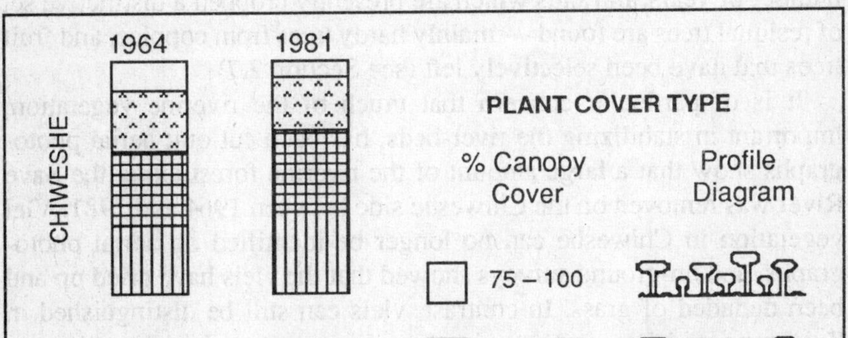

Data from aerial photographs.

As is to be expected, the least altered vegetation types in Chiweshe are the communities on the larger hills and mountains (i.e. the *Brachystegia glaucescens* community and the *Combretum apiculatum – Acacia nigrescens* community). On the smaller hills, the *Kirkia acuminata* community has in many cases been reduced from a woodland to a shrubland. The *Colophospermum mopane* community is still common in alluvial zones of Chiweshe, but it, too, has been converted to a shrubland. The major vegetation change in Chiweshe is on the plains where the *Julbernardia globiflora* community on the gneiss-derived soils and the *C. apiculatum – A. nigrescens* community on doleritic soils have been almost entirely cleared for cultivation. Only in remote areas are there a few stands of the former community, while the latter community can be found only on those soils which are rocky and shallow, and even then this community is much reduced in stature.

A number of derived plant communities are recognizable. One of these contains a large number of spinose, small trees (such as *Acacia nilotica* and *Dichrostachys cinerea*). It occurs in areas which were formerly cropped but which have been out of production for at least a number of years. In fields which are presently cropped a distinctive set of residual trees are found — mainly hardy trees from coppice, and fruit trees that have been selectively left (see Section 2.7).

It is of particular concern that much of the riverine vegetation, important in stabilizing the river-beds, has been cut out; aerial photographs show that a large amount of the riverine forest along the Save River was removed on the Chiweshe side between 1964 and 1981. Vlei vegetation in Chiweshe can no longer be identified on aerial photographs, and on-ground surveys showed that the vleis have dried up and been denuded of grass. In contrast, vleis can still be distinguished in Zviyambe and are common in the adjacent Romsley Resettlement Scheme.

3.3 Overgrazing

The question of overgrazing in Zimbabwe's CAs is a controversial topic not only because farmers are extremely sensitive to any suggestion of reducing herd sizes but also because the general scientific opinion has recently been challenged by Sandford (1982). On the basis of a highly critical review of literature on the Zimbabwean rangeland situation,

Sandford concluded that the available evidence does not support the common contention that environmental degradation has taken place. He defines environmental degradation as a decline in the productivity of natural resources (i.e. land and water resources) which is irreversible except at prohibitive cost. While conceding that soil erosion is taking place in CAs, and is probably causing a decline in productivity, he questions whether this constitutes environmental degradation and suggests that redeposited soil may be more useful in its new site than in the source area. Above all, he questions the link between livestock management in the CAs on the one hand and erosion, adverse hydrological changes, loss of primary productivity, etc., on the other.

These views have been echoed by some other researchers in Zimbabwe (such as Callear, 1984), and since they are of fundamental significance to future development policies they warrant discussion in the light of observations of the Chiweshe situation. Firstly, it has been shown that soil erosion from Chiweshe's cropland is causing serious decline in productivity (see Section 3.5), but models for the estimation of soil loss from rangeland have not been developed to suit Zimbabwean environmental conditions and therefore the rate of soil loss from grazing areas of Chiweshe was not quantified. There were, however, numerous signs of severe soil loss in such areas in the form of both sheet and gully erosion, at a scale that will require great investment of labour and/or machinery if the problem is to be prevented from worsening. A particularly difficult form of erosion to control in Chiweshe is the tunnel erosion that occurs in sodic soils in alluvial zones; the *Colophospermum mopane* community that grows in such zones (along drainage lines) has very sparse ground cover and the erosion hazard is further increased by the high volumes of runoff water that rush through the drainage lines during rainstorms.

Sandford regards the rate of soil erosion as being an indicator rather than a measure of environmental degradation. This is a matter of semantics in the case of Chiweshe: the erosion need not be quantified for its severity to be appreciated, and even if this degree of erosion only *indicates* environmental degradation then there is, none the less, cause for grave concern over future productivity of the grazing land.

The next issue to consider is the extent to which the erosion, and associated loss of primary productivity, can be ascribed to livestock. A cursory examination of a satellite image of the Save catchment is sufficient to reject Sandford's suggestion that any erosion that is occurring in the CAs may be due to slopes that are generally steeper than those of large-scale commercial farming areas, rather than to overstocking. The boundaries of the CAs of the catchment show as stark demarcations between these denuded areas and the adjacent commercial farming areas; clearly, natural environmental factors cannot be responsible for such regularity.

Considering possible management practices that could account for such a contrast in the spectral properties of the two tenure systems, the only plausible ones are wood-chopping, and livestock management. Wood-chopping is certainly a major factor as far as clearance of patches of arable land for cultivation is concerned, but less so with wood collection for fuel and construction since the Chiweshe residents do not observe tenurial boundaries when collecting wood (see Section 2.7). Also, felling of wood species for fuel and construction tends to yield more bushy vegetation than exists in the climax woodland because most woodland species coppice after being cut (see Section 3.2). Therefore, the bareness of grazing areas cannot generally be ascribed to wood-chopping. Overall, the conclusion is reached that the great contrast between the tenure systems is the result of the clearance of arable sites for cultivation and the drastic reduction of ground cover in the grazing areas by excessive numbers of cattle and goats.

It is certainly true, as Sandford states, that recommendations on stocking rates should take into account the fact that the CA farmers do not seek the same kind of commercial gain from their herds as large-scale ranchers do, and therefore the total numbers of animals that can be supported may be more important to them than beef production. However, the bottom line is the state of the environment, and it is felt that, on the basis of ecological observations made in the rangeland of Chiweshe both in the drought period and subsequently, it can be categorically stated that the environment is degraded and that this can be attributed largely to excess livestock. That the ecological carrying capacity has been exceeded is also shown by the recent drastic die-off of cattle (see Section 2.4).

The basic reasons for the problem of overgrazing are, as has been mentioned, that the multipurpose role of the cattle herd drives farmers to maximize the numbers of cattle that they own rather than their condition (see Behnke, 1985, for detailed discussion on this), and, secondly, that the grazing resource is a 'free-for-all'. It is a true 'tragedy of the commons' situation (Hardin, 1968), with each farmer attempting to obtain short-term, individual advantage from the resource at the cost of reduced long-term community benefit. As population increases, the grazing land (because of its communal ownership) is converted into individually-owned fields and is used for settlement, and the decreasing grazing area has to carry an increasing livestock population. The situation is aggravated by the fact that many households keep small ruminants as well as cattle. Although mixed ruminants can, generally, improve the productivity of the grazing land because of different grazing/browsing preferences, small ruminants (goats in particular) may be detrimental to the environment if they are allowed to roam uncontrolled on land that is already very denuded.

3.4 Hydrological changes

With the reduction in vegetation cover within the CAs of the upper Save catchment, it is to be expected that there would be a considerable alteration in the hydrological balance. A reduction in ground-water reserves was mentioned by several Chiweshe residents (see Section 3.1), and was invariably ascribed to diminishing rainfall over the last couple of decades. However, a study of rainfall records does not show any long-term trend towards lower mean annual rainfall in the upper Save catchment.

Such hydrological issues were researched by Du Toit *et al.* (1985), who analysed flow data from a gauging weir on the Save River, where the river runs between Chiweshe and Zviyambe, and rainfall data from the nearest meteorological station (Wedza). Multiple regression analysis revealed that estimated annual stream flow of the Save at this point was $2{,}53 \times 10^8$ m^3 greater in the late 1970s than in the mid-1950s, and the increase (which represents almost a third of the long-term mean annual flow) was independent of rainfall changes. Greater stream flow is probably because the loss of water from the catchment system due to evapotranspiration would be much reduced after large-scale removal of plant cover. Another likely consequence of vegetation degradation is a more rapid runoff response after rainfall, since water would not infiltrate readily; Bangezhano (1983) found that the rate at which water infiltrated into granitic soils of Chiweshe was one quarter of the infiltration rate in the adjacent Romsley Resettlement Scheme area, where soils are also granitic but deforestation and grazing pressure are much less. The short-term rainfall-runoff response could not, however, be investigated owing to a lack of data on daily stream flows.

A greater total stream flow of the Save, and a probable increase in the tendency of the river to flood after major storms, are of more significance to downstream irrigation schemes than to most of the Chiweshe residents. However, a significant reduction in the ground-water reserves due to impeded infiltration would obviously be a very serious local problem in this low-rainfall area.

3.5 Soil erosion and siltation

Observations on the extent of soil erosion in the grazing areas of Chiweshe have been included in Section 3.3. Soil loss from cultivated areas could be quantified through the use of the Soil Loss Estimation Model for Southern Africa (described by Elwell and Stocking, 1982). With this model, and taking into account average conditions of rainfall energy, soil erodibility, crop cover, slope percentage and slope length, Du Toit (1985) estimated minimum soil loss from the more gentle fields of Chiweshe to be of the order of 40 tonnes per ha per year. On steeper slopes (4 per cent or more), this soil loss would be doubled, and above-average rainfall would also greatly increase erosion, so that under some possible combinations of conditions the soil loss would be over 100 tonnes per ha per year. An appreciation of the significance of this can be gained by considering the decline in soil productivity resulting from even the lowest estimated rate of soil loss. Through the use of an equation developed by Elwell and Stocking (1984) to predict the remaining productive life of eroding soil, it was found that within twenty years (and in some localities within ten years) the Chiweshe soils will be too shallow to produce reasonable yields of maize. The productive life of the soil is even more reduced through the soil nutrient loss that accompanies erosion.

This high soil loss could be reduced considerably if more conservation works (erosion control terraces and storm drains) could be constructed in the fields, and if the practice of tied-ridging was adopted (Du Toit,

1985). Legislation introduced by the past Rhodesian government to force farmers to dig contour ditches was a source of great discontent, owing to the hard labour involved, and land husbandry became a highly political issue. The Zimbabwean government has not attempted to enforce any land husbandry measures, and Truscott and Pambirei (1983) reported that almost half of the Chiweshe households have inadequate terraces in their fields, or none at all. A broken terrace may, in fact, be a greater erosion hazard than none at all, since it would tend to funnel runoff and cause gullying. The potential effectiveness of well-constructed terraces (such as were observed on many Zviyambe farms) in controlling erosion is demonstrated by the relatively low rates of soil loss from commercial cropland. Meikle (1982) estimated the average rate of soil loss from granitic fields of 5 per cent slope in the large-scale commercial farming areas of the upper Save catchment to be only 6–9 tonnes per ha per year (this is also owing to the denser crop cover, preventing rainsplash).

Apart from the certainty that with present rates of erosion the soil resource will be too degraded to meet the increased agricultural requirements of the next generation in Chiweshe, there is also the problem of downstream siltation affecting national irrigation schemes. The construction of the proposed Chitowe reservoir on the Save, to make the Chisumbanje irrigation scheme one of the largest smallholder irrigation developments in Africa, has been delayed because the rate of siltation is thought to be excessive. At the Middle Save irrigation scheme, siltation of the river's channel has raised the level of water within the levees so that there is now a serious threat of overbank flooding. At both Chisumbanje and Middle Save, the life-span of irrigation pumps has been cut from seven years to three years because of the greater wear resulting from increased silt levels (according to the General Manager of the Agricultural and Rural Development Authority; Regional Water Authority, 1986).

Since the Save basin contains the largest tracts of irrigable soils in Zimbabwe (more irrigable land than can be irrigated with the available water yield of the Save), and is the major hope for future agricultural development in the country, siltation problems due to poor land husbandry upstream are of national importance. Sandford's (1982) suggestion that soil which is eroded in Zimbabwe's CAs may form useful deposits elsewhere is fallacious in this case.

3.6 Declining soil fertility

Residents of Chiweshe and Zviyambe were asked if food production was lower now than ten years ago, and 58 per cent of respondents said that this was the case; of these, 64 per cent cited soil exhaustion as the reason.

Mention has been made (Section 2.1) of the large areas which were formerly cultivated and which have now been abandoned owing to the depletion of soil nutrients. Grant (1981) has found that when nutrients in granitic soils are depleted to the extent that crops fail before the land is left fallow, the growth of nutrient-replenishing herbs and legumes is inhibited by the lack of phosphate and sulphate, and the poor ground cover leads to further leaching of minerals. Thus the lack of substantial regrowth in the abandoned fields is due not only to browsing by livestock but also to a loss of nutrients possibly beyond the level of medium-term recovery. The almost insurmountable problems associated with maintaining the fertility of cultivated fields, let alone rehabilitating abandoned land, have been discussed in Section 2.6.2. The best hope for meeting both requirements may well lie in agroforestry practices, with drought-resistant, nodulated legumes such as *Casuarina* being of great potential value. However, insufficient research has been done on agroforestry in Zimbabwe for firm recommendations to be made on suitable species and production systems.

It was of interest to find that Chiweshe farmers regarded the reddish dolerite-derived soils as being more productive than the coarser granitic soils in years of average rainfall, but felt that the latter were more productive in drought years. This is to be expected since the fersiallitic (moderately leached) granitic soils have a lower inherent fertility, but also a lower moisture tension, than the more clayey doleritic soils (Thompson and Purves, 1978).

3.7 Local attitudes on environmental change

Discussion with farmers in Chiweshe and Zviyambe revealed that in both areas, but more so in Chiweshe, the people were conscious of a trend of environmental degradation, and were prepared to ascribe many of the problems to poor land-use practices. However, the people tended to regard remedial action as being beyond their means (Du Toit, 1985).

Considering soil erosion, 47 per cent of respondents in Chiweshe stated that their fields had been damaged through erosion, and, of these,

91 per cent admitted that they could control the problem if they built contour ridges. On the question of overstocking (which had to be broached with great care owing to the lingering resentment over past compulsory destocking during the Rhodesia era), farmers in Chiweshe were about equally divided between those who believed that too many livestock damage the land and those who did not. In Zviyambe, 82 per cent of farmers felt that overstocking was ecologically detrimental. An indication of the extent to which the farmers believed that their areas were actually overstocked was gained by asking if there was sufficient grazing available. In Chiweshe, just over half the respondents (54 per cent) said grazing was insufficient, whereas just under half of Zviyambe respondents (47 per cent) said that this was the case. In both areas, destocking was suggested as a solution by a predictably low percentage (15 per cent in Chiweshe and 11 per cent in Zviyambe). The most commonly suggested solution was the provision of supplementary grazing areas elsewhere (on commercial farms or state land); this was cited by 38 per cent in Zviyambe and 52 per cent in Chiweshe. Suggestions were made by 24 per cent of the Chiweshe respondents for fenced grazing schemes within their area.

These local perceptions of environmental problems were generally encouraging, in that they revealed some awareness of relationships between land-use practices and ecological stability (to be expected with people so directly dependent on natural resources). Chavunduka (1982) came to similar conclusions, noting that there was a 'fairly sophisticated awareness of the relationship between land-use and ecology' (p. 289). However, some of the themes that emerged during the questionnaire survey and in discussions with farmers' and women's groups are disquieting for those concerned with ecological problems in Zimbabwe's farming areas. While people often accepted that their farming practices are degrading the land, they also believed that they should not be expected to do much about this since they are poor and short of labour and equipment; land exploitation appeared to them to be a right, and land conservation a luxury. It is clear that the farmers are trapped within various constraints; they do not 'destroy their land unless they have or can perceive no alternative' (Chavunduka, 1982, p. 292).

4 SOCIO-ECONOMIC ASPECTS

J. M. Clarke, J. A. Codd, B. H. Kinsey and R. H. Loewenson *

4.1 Infrastructure and services

The infrastructure in the CAs of the Save catchment has previously been described as inadequate (Truscott and Pambirei, 1983; Zimbabwe, 1983a). This is typical of the CAs in the country as a whole, since well-developed infrastructure is confined to the main watershed and to the intensive large-scale commercial farming areas of Mashonaland. This reflects past policy to avoid 'mutually harmful competitors' and to 'free' labour for other sectors (Yudelman, 1964) by the introduction of the Land Apportionment Act of 1930 where land was allocated by racial categories. The infrastructure operated exclusively in the interests of the owners of land and capital. The CAs themselves had to pay for the development of their infrastructure from levies on the livestock and crops. This put the CAs at an increasing disadvantage in relation to the well-developed areas where the improved infrastructure was reducing production costs. Any planning that was done in the CAs focused on agriculture only, and therefore the move to nucleated villages as formulated in the Land Husbandry Act of 1951 did not include the establishment of shops, schools, clinics and marketing facilities. The liberation struggle, too, has had its effects on the infrastructure: roads and bridges deteriorated owing to lack of maintenance, and cattle dip-tanks and small stores were destroyed.

Within the CA farming areas of the upper Save catchment there is a trend of decreasing development towards the south (Fig. 4.1). Development that did occur in these areas was in direct response to economic viability, with the areas in the higher rainfall regions receiving more attention.

4.1.1 *Transport*

The principal transport facilities in the Save catchment run through the

* Compiled by B. M. Campbell.

Figure 4.1: INFRASTRUCTURE OF THE UPPER SAVE CATCHMENT

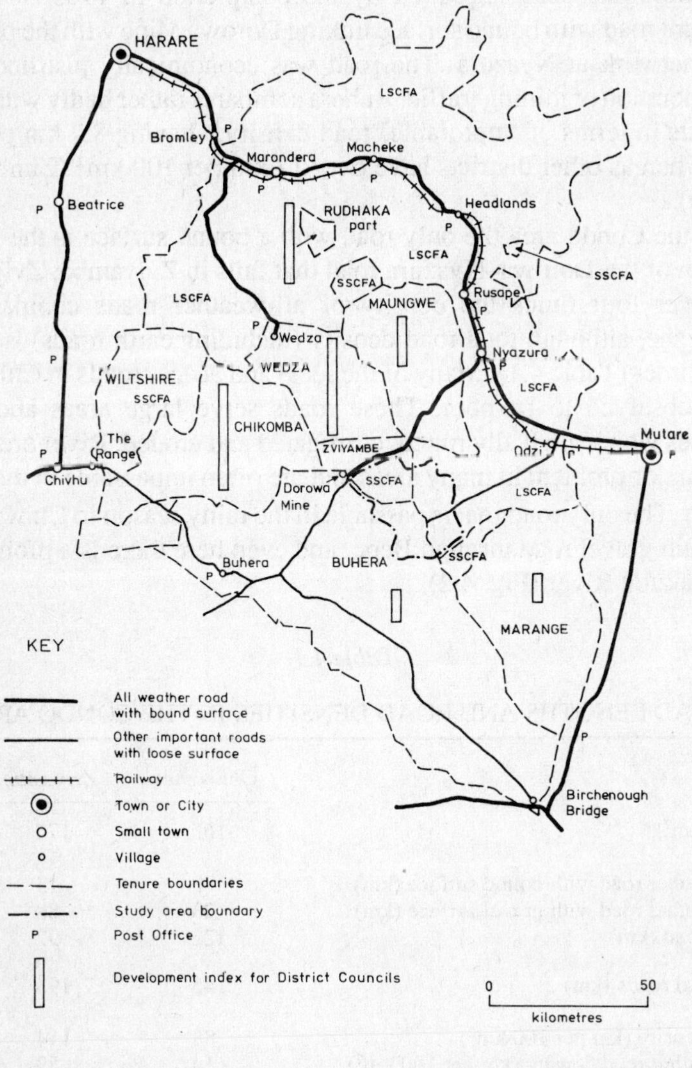

Also shown are development indices for District Councils (includes only CAs). The development index is derived from a principal components analysis of data on services and facilities.
Source: Zimbabwe (1982a).

large-scale commercial farming areas linking the main towns with Harare and Mutare by rail and road (Fig. 4.1). Access to Buhera and Zviyambe has been improved by the completion in 1983 of a high standard road with bound surface linking Dorowa Mine with the primary road network at Nyazura. The road was economically justified from consideration of mining traffic. Buhera compares rather badly with other districts in terms of 'motorable' road density,* having 8,7 km per 100 km² whereas other districts have over 17 km per 100 km² (Zimbabwe, 1982a).

In the Condo area the only road with a bound surface is the 13 km section of the Dorowa–Nyazura road that falls in Zviyambe. Zviyambe has over four times the density of all-weather roads compared to Chiweshe, although total road density (including earth roads) is of the same order (Table 4.1). Many of the local and access roads in Chiweshe were observed to be poor. These roads serve large areas and have surfaces which are badly rutted, corrugated and eroded. River crossings are a major problem in many areas and are often impassable in the rainy season. The only road that is passable in the rainy season in Chiweshe is the main gravel road through Bepe, and even here there is a problem at the Makubu River (Fig. 4.2).

Table 4.1

ROAD LENGTHS AND ROAD DENSITIES IN THE CONDO AREA

	Chiweshe	*Zviyambe*
Area (km^2)	165	170
All-weather road with bound surface (km)	0	13
All-weather road with gravel surface (km)	23	88
Earth road (km)	122	92
Total roads (km)	145	193
Road density (km per 100 km^2)	88	114
All weather road density (km per 100 km^2)	14	59

Data from aerial photographs.

* 'Motorable' road — all-weather, engineered gravel construction.

Figure 4.2: INFRASTRUCTURE IN THE CONDO AREA

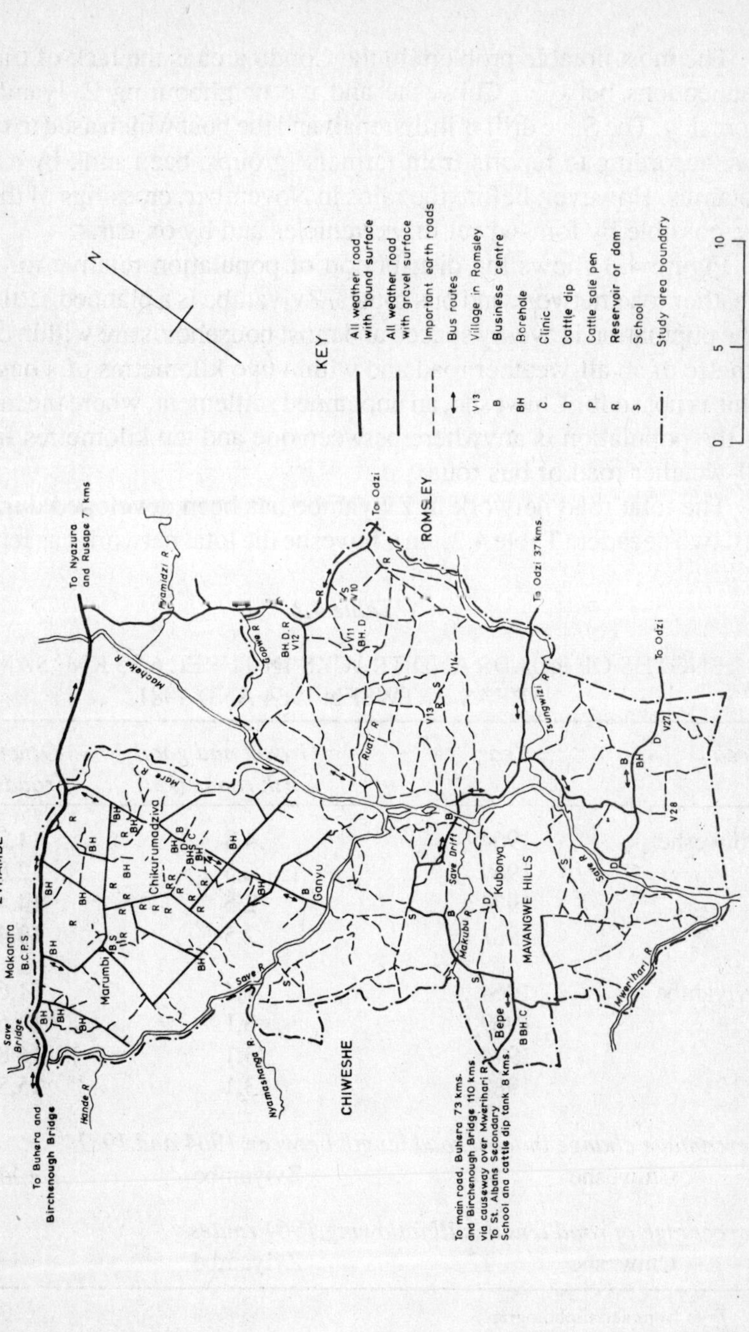

The most notable problem in the Condo area is the lack of transport connections between Chiweshe and the neighbouring Zviyambe and Romsley. The Save drift is in disrepair and the boat which used to operate has, according to reports from farmers' groups, been sunk by a hippopotamus. However, before the rains in November, crossings of the Save are possible by four-wheel-drive vehicles and by ox-carts.

Figure 4.3 shows the distribution of population relative to the all-weather road network and bus routes. Zviyambe is a planned settlement. The population is evenly spaced and most households are within one kilometre of an all-weather road and within two kilometres of a bus route. This is not so for Chiweshe, an unplanned settlement, where the majority of the population is anywhere between one and ten kilometres from an all-weather road or bus route.

The total road network in Zviyambe has been developed during the last two decades (Table 4.2). In Chiweshe the total network has remained

Table 4.2

LENGTHS OF ROADS AND TRACKS IN THREE 6,25 KM² SAMPLE AREAS BETWEEN 1964 AND 1981

Area	*Year*	*Gravel and good earth roads (km)*	*Other earth roads (km)*
Chiweshe	1964	4,8	11,2
	1969	2,5	12,8
	1974	2,5	14,4
	1981	2,5	13,4
Zviyambe	1964	1,4	11,6
	1969	3,1	14,6
	1974	3,1	14,8
	1981	3.1	15,5
Percentage change in total road length between 1964 and 1981			
Chiweshe	–1	Zviyambe	+44
Percentage of road length still following 1964 routes			
Chiweshe	47	Zviyambe	37

Data from aerial photographs.

Figure 4.3: POPULATION DISTRIBUTION IN THE CONDO AREA

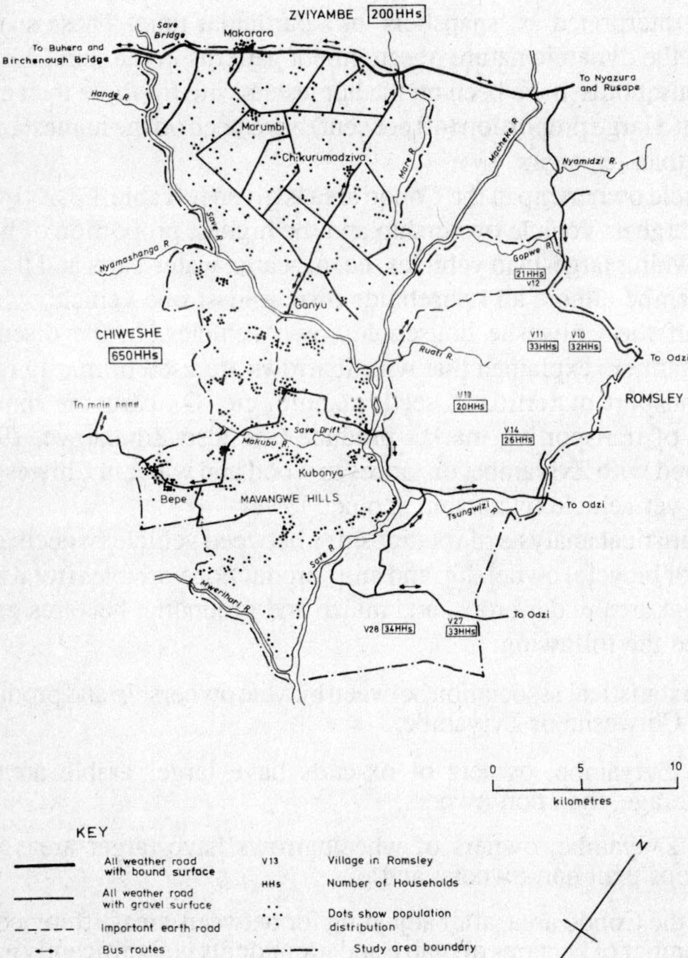

relatively constant although some sections of the road appear to have deteriorated in condition, presumably owing to lack of maintenance. Of interest is the fact that only 40 per cent of the roads in 1981 follow their 1964 routes, a phenomenon which has been observed in other CAs (Keech, 1969). In this context, statistics on earth roads and tracks can only be interpreted as 'snapshots' at a particular time. These statistics indicate the dynamic nature of settlement patterns in CAs; in Chiweshe most households have been on the homestead site for more than eleven years but a large proportion (40 per cent) have been on the homestead site for less than ten years.

Vehicle ownership in the Condo area is given in Table 4.3. Zviyambe has the highest vehicle ownership and the highest proportion of households owning large-load vehicles, i.e. ox-carts, water-carts and tractors. In Zviyambe almost all households own at least one vehicle, whereas only half the Chiweshe households own vehicles. In the discussion groups farmers explained that wheelbarrows are used primarily for on-farm transport of fertilizer, seed, manure, etc. Ox-carts are the main method of transporting market produce (see also Zimbabwe, 1983a). Compared with Zviyambe, distances to wood and water in Chiweshe are greater, yet vehicle ownership is much lower.

A statistical analysis of relationships between vehicle (wheelbarrow, ox-cart or bicycle) ownership and crop-production variables (total arable area, total area under crops, and maize and groundnut hectares grown) indicated the following:

(a) no statistical association between bicycle ownership and production in Chiweshe or Zviyambe;

(b) in Zviyambe, owners of ox-carts have larger arable areas, on average, than non-owners;

(c) in Zviyambe, owners of wheelbarrows have larger areas under crops than non-owners; and

(d) in the Condo area, after adjusting for between-area differences, the number of hectares of maize and groundnuts is significantly greater for owners of wheelbarrows (4 ha of maize, 2,4 ha of groundnuts) than for non-owners (1,6 ha of maize and 0,9 ha of groundnuts).

In short, the results suggest that vehicle ownership and agricultural production are not independent.

Table 4.3

OWNERSHIP OF VEHICLES, AGRICULTURAL EQUIPMENT AND RADIOS

	Chiweshe (n = 112)	*Zviyambe (n = 55)*	
Vehicles			
Ox-cart	30	89	p < 0,001
Water-cart	1	19	p < 0,001
Wheelbarrow	20	78	p < 0,001
Tractor	0	5	n.a.
Bicycle	35	69	p < 0,001
Any one of the above vehicles	51	95	p < 0,001
Agricultural equipment			
Ox-plough	83	98	0,001 < p < 0,01
Cultivator	20	91	p < 0,001
Harrow	13	78	p < 0,001
Ox-planter	1	28	p < 0,001
Cotton-sprayer	5	33	p < 0,001
Radio	28	43	0,01 < p < 0,05

Percentage of households.

All significant differences between areas indicated in this chapter were derived from chi-squared analyses.

n.a. = test not applicable because expected frequencies too low; d.f. = 1 in all cases.

4.1.2 *Agricultural facilities and services*

There are marked differences between Chiweshe and Zviyambe in terms of access to agricultural facilities and services such as markets, extension, veterinary service and agricultural equipment.

The proportion of farmers using credit in Zviyambe is slightly higher than that in Chiweshe, but, in general, the use of credit facilities is not widespread in the Condo area (Table 4.4). What the data do not tell us is whether all farmers who wish to use credit are, in fact, doing so. Until the early 1980s, there has been no national programme of credit for farmers in the CAs, and the recently introduced scheme is still in the very early stages of expansion. Access to credit should become easier in Chiweshe.

However, it could be anticipated that private sources of credit, and even parastatals primarily orientated towards the conditions and demands of large-scale agriculture, may not operate on terms favourable to CA farmers, so that while credit is apparently available, conditions governing access to it may exclude many farmers. Farmers in SSCFAs, such as Zviyambe, have always had reasonably good credit services.

Table 4.4

ACCESS TO AGRICULTURAL CREDIT AND EXTENSION ADVICE

	Chiweshe (n = 113)	*Zviyambe (n = 58)*	
Receiving credit for the 1982/3 season	4	12	n.s.
Extension visits at least monthly	12	56	$p < 0{,}001$
	(n = 93)	*(n = 57)*	
Visits by veterinary department personnel to households with cattle	67	90	$0{,}001 < p < 0{,}01$

Percentage of households.
d.f. = 1 in all cases.

Zviyambe farmers enjoy a much higher level of on-farm extension visits, 85 per cent receiving visits (mostly monthly) compared with only some 40 per cent in Chiweshe (see also Table 4.4). General discontent with access to agricultural extension advisers was expressed in Chiweshe.

Most households own cattle and since their cattle serve multiple uses, poor animal health and high mortality can have devastating consequences for the household economy. An effective veterinary service, through inoculation, treatment and advisory programmes, can do much to ensure the health of the household herds. Table 4.4 indicates that the on-farm veterinary services are reasonably effective, more so in Zviyambe, in providing coverage to cattle-owners. The situation with cattle-dip facilities is, however, problematical (Fig. 4.2). The farmers' groups in Chiweshe reported that cattle-dips were too few and too far for cattle to travel to. Distances to the dip at Kubonya in Chiweshe are up to fourteen kilometres for some households; a round trip, in these cases, could take twelve hours. We suggest that the maximum round-trip travel time should be in the order of four hours, excluding dipping time.

Zviyambe farmers are also at an advantage with respect to agricultural equipment (Table 4.3). Of all the agricultural equipment, it is only ox-ploughs that are frequently owned by Chiweshe farmers. In general, there are five-fold higher ownership levels in Zviyambe of cultivators, harrows, ox-planters and cotton-sprayers.

Much dissatisfaction was expressed with marketing facilities in both Zviyambe and Chiweshe, although Chiweshe farmers are clearly in the more dire situation. There is a cattle sales-pen at Makarara Business Centre in Zviyambe, but no sales-pen exists in Chiweshe. Table 4.5 indicates that whereas Zviyambe farmers rely on formal marketing outlets — the Grain Marketing Board and marketing co-operatives — for their crop sales, Chiweshe farmers often rely on local traders who, unlike the Grain Marketing Board, do not have fixed prices. Accessibility to marketing facilities is hampered by large distances to existing facilities, difficulties in marketing small loads, dependency on and the high cost of hired trucks, and the poor condition of the roads.

Table 4.5

MARKETING OUTLETS FOR MAIZE AND GROUNDNUT GROWERS

	Chiweshe		*Zviyambe*	
Marketing outlet	*Maize growers (n = 36)*	*Groundnut growers (n = 14)*	*Maize growers (n = 21)*	*Groundnut growers (n = 22)*
GMB depot	30	23	37	35
GMB agent	8	8	26	30
Marketing co-operative	8	8	37	35
Traders	38	55	0	0
Other farmers	8	8	0	0
Non-local sources	8	0	0	0

Percentage of growers that marketed produce after the 1982/3 season. Sample sizes too small for chi-squared analyses..

4.1.3 *Water and sanitation*

Access to a clean and reliable supply of water for domestic use is a major determinant of health status, yet more than half the households in Chiweshe rely on surface supplies of water (rivers, springs) for domestic use (Table 4.6). Access to these worsened during the drought as many sources dried up and the remaining ones experienced heavier rates of utilization. The next most common sources of water for domestic purposes in Chiweshe (the major sources in Zviyambe) are boreholes and pumps. Provided a borehole is adequately constructed with proper above-ground waste-water disposal, it is a safe source of potable water; and the same can be said for wells with pumps. Unprotected wells, however, while better than surface supplies, are prone to contamination. Taps for individual households are very rare and occur only in Zviyambe.

Table 4.6

NATURE OF DOMESTIC WATER SUPPLY

Water supply	*Chiweshe (n = 108)*	*Zviyambe (n = 57)*
Borehole and pump	30	58
Well and hand-pump	7	9
Unprotected well	11	6
River, spring	52	27

Percentage of households.
There is a significant difference between areas ($0,001 < p < 0,01$; d.f. = 3).

Distances to water are usually considerable and Table 4.7 indicates that some 40 per cent of households have to travel more than four kilometres per day for water; only about a quarter of households walk less than one kilometre per day. In Kenya, the target for 1983 was to have more than 85 per cent of people within two kilometres of a water source (Ghai *et al.*, 1980). The most common way of fetching water is for women to walk to the source and carry water home in head-borne 20-litre containers, a pattern familiar to many underdeveloped countries. The large majority of households transport between 40 and 60 litres per day, in two or three trips (although up to five trips per day were observed in

Table 4.7

DISTANCE WALKED TO COLLECT WATER

	Chiweshe (n = 110)	*Zviyambe (n = 54)*
0–500 m	21	12
500–1000 m	9	9
1–2 km	13	15
2–4 km	16	11
4–8 km	25	28
8–32 km	17	26

Percentage of households.
No significant difference between areas (n.s., d.f. = 5).

the surveys). Water consumption falls well short of Unicef's recommended 38 litres of household water per person per day (actual water consumption per person is difficult to calculate as many people use the Save River for washing themselves and their clothes).

The data in Table 4.8 show that Chiweshe has a long way to go in terms of improved domestic and community hygiene. Over 80 per cent of households have no facilities for the safe disposal of human waste.

Table 4.8

DOMESTIC SANITATION FACILITIES

	Chiweshe (n = 112)	*Zviyambe (n = 58)*
Human waste p < 0,001		
Private, ventilated pit-latrine	14	59
Communal pit-latrine	5	0
None	82	41
Household waste 0,01 < p < 0,05		
Refuse pits	74	89
None	26	11

d.f. = 1 in all cases.

Historically, when population pressures were very much lower than they are today, this fact need not have been of great concern, for biological degradation of waste could occur with minimal danger of contact or contamination. However, population density in Chiweshe is now high and, in addition, people are still relying mostly on unprotected surface-water supplies — a situation where risk of water-borne and fly-borne diseases is high. The risk of these diseases is likely to be lowest in Zviyambe, where the majority of families have pit-latrines and use borehole water. Most households have refuse pits (Table 4.8).

The survey shows an even greater reliance on surface-water sources for watering livestock. Our data do not confirm the fact, but it is likely that most surface-water sources are being used for both livestock and domestic purposes, thus compounding the health hazard. Livestock have to walk, on average, 1,9 km in Zviyambe and 2,1 km in Chiweshe to water (even 9 km in some cases). In Zviyambe almost all farms have at least one small earth dam, which further improves the water situation for Zviyambe farmers. However, during the drought, these dams were empty.

4.1.4 *Housing*

Perhaps more than any other basic need except food, shelter reflects the environment in which it is found. The environment both provides the building materials and shapes the way they are utilized to provide shelter from environmental extremes. Table 4.9 employs an ordinal hedonic scale to display the types of shelter found in the sample. The scale ranks types of shelter from most to least preferred and was constructed by having respondents rank their preferences.

More than 60 per cent of the Chiweshe sample live in the two least-preferred shelter types — those constructed entirely from locally gathered materials. In Zviyambe only 23 per cent of households live in these types, with the most common types being the two most-preferred types, a reflection of the much higher incomes to be found in Zviyambe (see Section 4.2.2). We can deduce that households in Zviyambe use their higher incomes to 'import' a fair amount of purchased housing material. It appears that the adequacy of the resource base (trees for construction and grass for thatching — see Chapter 2) is not always the dominant influence on type of shelter used, because Zviyambe households have better access to these resources and yet it is in Zviyambe that households use less of the locally gathered resources.

Table 4.9

CHARACTERISTIC FEATURES OF FAMILY'S SHELTER

	Chiweshe (*n* = *113*)	*Zviyambe* (*n* = *58*)
1. Burned brick and corrugated iron or asbestos roof with framed and glassed windows	16	26
2. Sun-dried brick and corrugated iron or asbestos roof with framed and glassed windows	8	19
3. Sun-dried brick and corrugated iron or asbestos roof without framed and glassed windows	6	18
4. Sun-dried brick with thatched roof and framed and glassed windows	3	14
5. Sun-dried brick with thatched roof without framed and glassed windows	44	16
6. Wattle and daub with a thatched roof	23	7

Percentage of households living in different types of shelter.
1 = most preferred; 6 = least preferred.
There is a significant difference between areas ($p < 0{,}001$; d.f. = 5).

4.1.5 *Education*

Table 4.10 presents data on school attendance for two age groups. In both the generation of young schoolchildren and their parents' generation there is a better-educated population in Zviyambe: relatively fewer people have no education and many more people have 5–12 years of education. There is a greater difference between Zviyambe and Chiweshe in the parents' generation than in the younger generation, indicating that the education policy of the new government is reaching the CAs.

Primary-school locations are generally satisfactory but access to the nearest secondary school in Chiweshe is problematical, with some children living ten to twenty kilometres from the school. School buses or bicycles could alleviate the problem, but both solutions emphasize the need for a cash income. It is considered that secondary-school pupils should not have to walk more than five kilometres and primary-school pupils more than three kilometres to school (Gould, 1978).

Non-formal education, in terms of agricultural extension, has been dealt with in Section 4.1.2. Table 4.3 indicates that Zviyambe households have greater access to radio than Chiweshe households, another source of non-formal education.

Table 4.10

PRIMARY SCHOOL ATTENDANCE FOR SELECTED AGE GROUPS OF THE SELF-DEFINED HOUSEHOLD

	Age group			
	6–14 years old		*35–50 years old*	
Years of schooling	*Chiweshe (n = 236)*	*Zviyambe (n = 144)*	*Chiweshe (n = 114)*	*Zviyambe (n = 115)*
No schooling	13	8	15	1
Four years	74	63	32	25
More than four years	13	29	53	74

Percentage of individuals.

There are significant differences between both areas for both age groups (p < 0,001; d.f. = 5 for each group).

4.2 Employment and income

4.2.1 *Employment*

Every household had at least one member who was occupied in farming. However, of the working-age members of the 'self-defined' household more than 30 per cent were involved in off-farm employment, mostly as migrant labourers (Table 4.11). Less than 10 per cent were employed locally, although many of those recorded as farmers do have an income from informal activities such as beer-brewing, handicrafts, etc. It is chiefly the males who have off-farm employment. Overall, Zviyambe has the slightly higher level of off-farm employment, but because of differences in average household size between Chiweshe and Zviyambe (Table 6.1) it is Chiweshe households which have fewer farmers per

Table 4.11

EMPLOYMENT STATUS OF WORKING-AGE MEN AND WOMEN AND ABSENTEEISM OF HEAD-OF-HOUSEHOLDS

	Chiweshe		*Zviyambe*	
	Men and women (n = 287)	*Women only (n = 122)*	*Men and women (n = 233)*	*Women only (n = 97)*
Employment [a]				
Farming	66	92	57	76
Local employment	7	2	3	1
Migrant labour	27	6	40	23

Head-of-households absent for more than 6 months per year [b]

Chiweshe (n = 113)	*Zviyambe (n = 58)*
25	7

[a] Percentage of individuals (men and women: 0,001 < p < 0,01; d.f. = 2; women only: p < 0,001; d.f = 1).

[b] Percentage of households (0,01 < p < 0,05; d.f. = 1).

save 5

household and which have a very high proportion of absentee heads-of-households (Table 4.11). Cross-tabulations of schooling and employment for both Chiweshe and Zviyambe indicate that off-farm employment is associated with 8–11 years of schooling, whereas farmers generally have fewer years of schooling.

4.2.2 *Income*

A family with an 'adequate' income will rarely have difficulty in meeting its basic needs; indeed, the very notion of adequacy of income, as enshrined in such concepts as the poverty datum line, is that the income should suffice to procure a family's basic needs. Table 4.12 shows that 93 per cent of Zviyambe households rely on rural-based primary production as their major income source. In Chiweshe it is only 47 per cent of households that rely on this income source, many households relying on urban-based income sources. Another important income source in Chiweshe is rural-based secondary production — principally handicrafts. Produce yielded from agriculture is clearly insufficient as a sole

Table 4.12

MAJOR SOURCE OF INCOME

	Chiweshe (n = 113)	*Zviyambe (n = 58)*	
Nature of primary income source			
Rural employment and trading	0	9	n.a.
Urban-based income sources	5	33	$p < 0{,}001$
Non-producers (salary-earners, e.g. teachers)	2	0	n.a.
Rural-based secondary production (mostly handicrafts, some beer-brewing)	0	10	n.a.
Rural-based primary production (mostly crops, some livestock)	47	93	$p < 0{,}001$
Households receiving regular remittances			
Food	13	12	n.s.
Money	20	12	n.s.

Percentage of households, as cited by respondents.
d.f. = 1 in all cases.

source of income, with only 40 per cent of Zviyambe households and only 22 per cent of Chiweshe households relying on agriculture as the sole income source.

Technosynesis (1984) estimates that average per capita annual incomes in Chiweshe are about Z$60* (an average from a drought year and a wet year). Truscott and Pambirei (1983) indicate that, for many households, the combined income from remittances and the informal sector make up more than the total income from agriculture. However, it should also be noted that very few households stated that they receive remittances (food or money) regularly (Table 4.12). The Z$60 per capita annual income gives some indication of the incredibly low levels of income rural households in CAs must survive on. We estimate that annual incomes in Zviyambe are at least Z$200 per capita because of the usually greater crop sales and the higher off-take of livestock.

The above patterns of employment and income are widespread in the CAs (see, for example, Callear, 1984; German Agency for Technical Co-operation, 1985).

4.3 Nutrition and health

4.3.1 *Food supply*

The majority of children under five in both Zviyambe and Chiweshe eat high-bulk (maize) meals three to four times a day, usually supplemented with vegetables at one of the meals. The diet in Zviyambe includes slightly more frequent consumption of high-bulk food, oils, vegetables and fruit, and considerably more frequent consumption of protein and groundnuts (Table 4.13). Patterns of consumption for the other household groups we investigated, schoolchildren and adult women, were similar. Small children do not have stomachs large enough to absorb in a day the calories they require from a bulky food such as maize, and thus concentrated foods with a higher oil content, such as groundnuts or beans, should be an important dietary component. In Chiweshe there is minimal daily consumption of oils and beans. In the case of calorie deficiency other foods are broken down to get the necessary energy requirement, therefore a malnourished child is often deficient in proteins as well as calories (Morley and Hauman, 1979; Berg, 1981).

* Z$1,00 = US$0,60 (1985), US$0,50 (1989).

Table 4.13

FOOD CONSUMPTION PATTERNS
OF CHILDREN UNDER FIVE YEARS OLD

		Chiweshe (*n* = 85)	Zviyambe (*n* = 43)	
Food consumption in the past 24 hrs	*Number of times eaten*			
High bulk				n.s.; d.f. = 1
(mostly maize-meal	1–2	19	12	
porridge, some bread)	3–4	81	88	
Oil, beans				n.s.; d.f. = 1
	0	81	77	
	1–3	19	23	
Fruit, vegetables				n.s.; d.f. = 2
	0	19	7	
	1–2	71	82	
	3–4	10	11	
Protein				p < 0,001; d.f. = 1
	0	82	53	
	1–3	18	47	

		Chiweshe (*n* = 94)	Zviyambe (*n* = 46)	
Frequency of consumption of:				
Groundnuts				p < 0,001; d.f. = 1
	never	48	15	
	sometimes	35	61	
	weekly	10	20	
	daily	7	4	
Oil, margarine				n.s.; d.f. = 2
	never	13	2	
	sometimes	41	35	
	weekly	12	13	
	daily	34	50	

Percentage of households with children under five.

Food supply during the period of survey was strongly linked to the drought-relief programme which, according to respondents, reached 86 per cent of the Zviyambe households, but only 64 per cent of the Chiweshe households. Presumably the Zviyambe settlement pattern with its good road network and clearly demarcated farms, ensured the efficient distribution of food aid. Most households receiving aid got maize and groundnuts or beans. A further roughly similar amount of maize was being purchased by households at the time of the survey. Most of the fruit and vegetables were locally grown, and the meat was mostly purchased, although a third of the meat consumed was from on-farm slaughterings. Vegetable gardens are more common, and there are more slaughterings, in Zviyambe than in Chiweshe (see Chapter 2). A considerable amount of food was bought by households. Maize meal accounted for the highest proportion of purchases: for the month prior to the survey, Zviyambe households spent an average of Z$16 on maize meal, whereas Chiweshe households spent Z$13. The next most important purchases were bread, oil and meat, with each household spending, on average, between Z$3 and Z$6 on each of these items. At the time of the survey Zviyambe households were spending Z$37 per month on food whereas Chiweshe households were spending Z$30 per month on food. Zviyambe, therefore, has better food supplies in terms of all three sources: drought relief, purchases and farm produce.

4.3.2 *Nutritional status*

Table 4.14 indicates that the nutritional status of under-five-year-olds is better in Zviyambe than in Chiweshe. For instance, protein energy malnutrition (PEM) as measured by weight-for-age indicates that some 50 per cent of children in Chiweshe show some degree of PEM, whereas about 30 per cent of children in Zviyambe show some degree of PEM. The cases of malnutrition in Zviyambe and Chiweshe are of a moderate and mild nature, with no cases of severe PEM being recorded. The weight-for-height parameter reflects wasting or acute malnutrition and is a measure of the more immediate circumstances of the child's health and food supply; 20 per cent of children in Chiweshe and 12 per cent in Zviyambe show some signs of wasting. The height-for-age parameter reflects stunting or chronic undernutrition and is a measure of the long-term effects of food supply and disease; 85 per cent of Chiweshe children and 70 per cent of Zviyambe children show some sign of stunting. A

Table 4.14

THE NUTRITIONAL STATUS OF CHILDREN UNDER FIVE YEARS OLD ACCORDING TO WHO (1978) STANDARDS

		Chiweshe (*n* = *148*)	*Zviyambe* (*n* = *87*)
Weight-for-age (PEM)	0,001 < p < 0,01		
Normal		51	71
Mild (80–90% of standard)		25	15
Moderate and severe (less than 80% of standard)		24	14
Weight-for-height (wasting)	n.s.		
Normal		80	88
Mild (85–90% of standard)		9	6
Moderate and severe (less than 85% of standard)		11	6
Height-for-age (stunting)	0,01 < p < 0,05		
Normal		15	30
Mild (90–100% of standard)		57	53
Moderate and severe (less than 90% of standard)		28	17

Percentage of children; d.f. = 2 in all cases.

quarter of the children in Chiweshe are moderately stunted, but less than 3 per cent are severely stunted. As is usual, in both areas stunting has a higher prevalence than wasting. Cross-tabulations of nutritional parameters and age indicate that wasting and stunting in Chiweshe is highest in the 13–36-month age group, which may be explained by early weaning practices and the lack of high-energy, low-bulk foods in the diet as well as the prevalence of diarrhoeal, respiratory and other communicable diseases in that age group (Chikanza *et al.*, 1981; see also Kenya, 1983; Unicef, 1984).

4.3.3 *Infant mortality rate*

Because of small sample sizes we are unable to estimate the infant mortality rate for Chiweshe and Zviyambe separately, but, overall, the infant mortality rate is 60 per 1 000 live births, which is relatively low compared with other estimates from underdeveloped countries. The sample may have underestimated infant mortality as this statistic was based on recall (see Section 6.2.1 for methodology). At a national level it has been found that more than 80 per cent of infant deaths originate in the perinatal period through communicable diseases, nutritional deficiencies and diarrhoeal diseases (Ministry of Health, pers. comm.). These mortality factors are all associated with poverty: poor hygiene, poor nutrition, inadequate sanitation, unsafe drinking water, lack of medical facilities and ignorance of basic preventive health care.

4.3.4 *Reported health problems*

During the survey over 30 per cent of women said that they had been ill in the last month, or were ill at the time. As women form the majority of the workforce, this illness rate must considerably affect the household. The most common complaint was that of general body pains (Table 4.15), a reflection of the long hours of physical labour and the long distances walked. The strain of these activities may be heightened by

Table 4.15

NATURE OF WOMEN'S LAST ILLNESS

Illness	*Chiweshe (n = 99)*	*Zviyambe (n = 51)*
General body pains	68	69
Heart, chest	12	12
Gynaecological/reproductive problems	7	10
Other (coughing, diarrhoea, accidents)	13	9

As reported by female respondents.
There is no significant difference between areas (n.s.; d.f. = 3).

undernutrition, frequent childbirth (average of 32 months between birth and conception) and parasitic and communicable diseases such as bilharzia, malaria and respiratory infection. At least 40 per cent of children in Save CA suffer from bilharzia, and the upper Save catchment is subject to epidemic outbreaks of malaria.*

The reported incidence of diarrhoea among children is high in both Chiweshe and Zviyambe. Over a third of families with children under

* Dr P. Taylor, Blair Research Laboratory, Harare, pers. comm.

five reported one or more cases of diarrhoea during the past month, with a slightly higher incidence in Chiweshe (43 per cent of Chiweshe children and 35 per cent of Zviyambe children). The incidence of diarrhoea was highest in the 13–24-month age group, a condition that may contribute to the high prevalence of wasting and stunting in the 13–36-month age group.

4.3.5 *Health services*

The two clinics in Zviyambe service about 1 500 people each, whereas the single clinic in Chiweshe services over 9 000 people. This situation is reflected in the data on frequency of clinic visits — 45 per cent of women in Chiweshe did not visit the clinic in the last year, whereas only 25 per cent of women in Zviyambe did not visit the clinic in the last year. Another health service is the village health worker scheme. Village health workers have an important role in promoting primary health care; however, their visits to households are infrequent: 58 per cent of Chiweshe households and 78 per cent of Zviyambe households had received no visit during the last year prior to the survey (the scheme is chiefly focused on the CAs).

The availability of health care is also indicated by health cards and immunization status. Health cards are given to mothers for monitoring the health of their children. Only 28 per cent of Chiweshe children under five years old had up-to-date health cards;this compared with 44 per cent in Zviyambe. In both areas there is a very low percentage of children who have been fully immunized (less than 10 per cent).

In general, it can be seen that much improvement is needed in health services, especially in Chiweshe.

4.4 Community organization

In general, the people in Zviyambe have a slightly greater membership of community organizations (Table 4.16), but the patterns of membership are quite different. Chiweshe heads-of-households are more involved in political organizations and religious associations and have very limited membership of farmers' groups, whereas Zviyambe heads-of-households are well represented in farmers' groups. More women than men are members of community organizations mainly because of the presence of women's groups. Women are much less represented than men in political organizations and farmers' groups. However, the women's groups do address themselves to farming matters. Perhaps the most striking difference between Zviyambe and Chiweshe is the four-fold higher membership of farmer's groups by Zviyambe heads-of-households. Furthermore, the Zviyambe farmers' organizations are relatively formal groups (Intensive Conservation Area group, marketing co-operatives) whereas the Chiweshe groups tend to be informal. The Zviyambe farmers' organizations undoubtedly play a role in promoting higher productivity through improved agroservices and informal education.

Table 4.16

MEMBERSHIP OF COMMUNITY ORGANIZATIONS

	Head-of-household (male or female)		*Women interviewed*	
	Chiweshe (n = 113)	*Zviyambe (n = 58)*	*Chiweshe (n = 113)*	*Zviyambe (n – 58)*
Percentage of households belonging to:				
No organization	32	26	23	19
Political organization	45	40	17	9
Religious associations	27	9	29	7
Farming organizations (farmers' groups, development committees)	13	55	2	14
Women's clubs	6	7	56	74
Others (village, school administration)	3	7	1	0
	p < 0,001; d.f. = 3		0,001 < p < 0,01; d.f. = 3	

4.5 Community identification of basic needs deficiencies

One of the questions on the questionnaire asked both the head-of-household and a wife to list the three most important facilities that they felt were needed in their area (Table 4.17). Health facilities, water supply and transport are clearly felt to be the most deficient facilities in the area, accounting for over 50 per cent of the needs expressed. Much of the emphasis given to water undoubtedly stems from the drought, but water supply would certainly remain a high priority even under more normal

Table 4.17

BASIC NEEDS DEFICIENCIES

	Chiweshe		*Zviyambe*	
	Household-head (n = 611)	*Wife (n = 612)*	*Household-head (n = 302)*	*Wife (n = 286)*
Deficiency/felt need				
Hospital or clinic	28	26	9	24
General water supply	17	24	39	38
Improved transport and marketing (mostly roads/bridges, also bus services)	26	28	13	6
Education (mostly related to secondary school)	8	7	7	6
Food aid/drought relief	4	2	7	6
Improved local shopping	6	8	3	1
Co-operative agricultural equipment	0	0	6	4
Maize-grinding mill	2	2	0	0
Seed	2	0	2	2
All other needs named	7	3	14	13

Percentage of responses.

As perceived by the head-of-household (man or women) and interviewed wife.

There are significant differences between areas in the distribution of answers in both the head-of-household and wife (both $p < 0{,}001$, with d.f. = 7 and d.f. = 6, respectively).

rainfall conditions. In Zviyambe, households placed considerably less stress on transport and slightly less on health, with water being the overwhelming problem. This is, of course, related to the adequate transport network in Zviyambe and the better provision of health services. Water is perceived as the main problem in Zviyambe even though Zviyambe is better provided for than Chiweshe in terms of boreholes and pumps (see Table 4.6).

Other important deficiencies related to education (particularly secondary schooling), drought relief and local shopping facilities. Emphasis on local shopping was more marked in Chiweshe where, unlike in Zviyambe, the business centres are widely scattered (Fig. 4.2) and poorly stocked. This reflects the little cash that is in circulation in Chiweshe. Many of the other needs expressed related directly to farming, but it is perhaps surprising that farming needs did not feature more prominently.

In Chiweshe these needs included improved accessibility to seed, grinding mills, pesticides, credit and extension. Many of the minor needs expressed by Zviyambe households concerned items which they were well provided with relative to Chiweshe households, such as agricultural equipment, improved sanitation, extension, fencing, roofing material and more land for grazing. Very few needs were related to natural resources other than water. Firewood and thatching grass were mentioned by a few, but, in general, infrastructural needs dominated. Furthermore, very few needs related directly to food and nutrition.

Table 4.18 identifies the major problems in farming as perceived by the heads-of-household. Once again, water — the unpredictable and low rainfall — is recognized as a major limiting factor but the other major problem is insufficient draught power. Water is seen as more of a problem in Zviyambe, perhaps because other constraints have been somewhat overcome. Another important problem, especially in Chiweshe, is the inadequate accessibility to seeds.

Table 4.18

PROBLEMS IN FARMING

	Chiweshe (n = 113)	*Zviyambe (n = 58)*
Insufficient draught power	21	29
Unpredictable/low rainfall	17	29
Inadequate/expensive seed	15	9
Insufficient labour	10	10
Inadequate transport	7	5
Inadequate implements	7	3
Problems with wild animals	4	3
Livestock diseases	4	3
Others (capital, weeds, extension, grazing area, erosion)	15	9

Percentage of responses, as perceived by the head-of-household.
No significant differences between areas (d.f. = 4).

5 LINKS BETWEEN THE ENVIRONMENT AND BASIC NEEDS SATISFACTION

B. M. CAMPBELL, M. J. SWIFT (COMPILERS)

5.1 Introduction

The objective of this chapter is to examine the linkages between the environment and the satisfaction of basic food and health needs. The conceptual framework for considering the interactions between environmental components and needs components has been dealt with in detail by McHale and McHale (1978); we thus restrict ourselves to analysing interactions within the study area, using the data presented in the previous chapters. In this analysis, we identify the key interactions in order to propose development policies which will promote the sustainable satisfaction of basic food and health needs (see Chapter 7).

5.2 Major features of the environment and needs satisfaction

The two ténure systems examined in depth show many differences both with respect to environment and satisfaction of food and health needs (Table 5.1). Clearly, health and food needs satisfaction is at a very low level in Chiweshe; there was a high prevalence of undernutrition and diarrhoea among children in Chiweshe. Zviyambe households, although only a few kilometres from their Chiweshe counterparts, have a much higher degree of needs satisfaction. An important difference between Chiweshe and Zviyambe relates to food intake, and it is significant that this difference was detected even after a drought year when the usually more productive Zviyambe farmers recorded very poor yields; in Zviyambe, despite the drought, there were higher food intakes, especially of protein and groundnuts.

It is impossible to gauge the change in needs satisfaction in the last decade. In the 1970s the population of Save CA grew at about 3 per cent per annum, and from the little agricultural data available for this period there is no indication that total production increased. Therefore, we can surmise that per capita food production has, if anything, decreased.

Table 5.1

DIFFERENCES BETWEEN CHIWESHE AND ZVIYAMBE IN 'NEEDS' COMPONENTS AND ENVIRONMENTAL COMPONENTS

	Chiweshe	*Zviyambe*	*Original data*
Indicators of food and health needs satisfaction			
Undernutrition	very high	high	Table 4.14
Frequency of diarrhoea	very high	high	Section 4.3.4
Food intake	low	higher	Table 4.13
Indicators of environmental status			
Deforestation	severe	less severe	Fig. 3.1
Erosion	severe	less severe	Sections 3.3 and 3.5
Area of grazing land per livestock unit	very small	small	Sections 2.4 and 3.3
Area of cleared, abandoned land	large	small	Sections 2.1 and 3.6
Population densities (people per km²)	29	16	Table 6.4

Many environmental differences have also arisen between Zviyambe and Chiweshe. In Chiweshe there is more severe degradation of the soil, woodland and grazing resources. Grazing land is in shorter supply, there are large areas of unproductive, cleared land, and per capita cropland is more limited.

5.3 Natural resources and infrastructure/services as determinants of needs satisfaction

Despite the clear differences in environmental components and needs satisfaction between Chiweshe and Zviyambe, it is difficult to unravel the linkages, to separate causes from effects, and to separate correlative factors from causative.

An important difference between Zviyambe and Chiweshe concerns infrastructure and services; Zviyambe is much better serviced with roads, schools, cattle-dips, markets, credit facilities and extension staff (see Table 5.2). The question thus arises: To what extent are the natural resources — soil, land, water and plant cover — responsible for the satisfaction or non-satisfaction of basic food and health needs, and how

Table 5.2

DIFFERENCES BETWEEN CHIWESHE AND ZVIYAMBE IN INFRASTRUCTURE AND SERVICES

	Chiweshe	*Zviyambe*	*Original data*
Factors affecting agricultural production and marketing			
All-weather roads	few	very many	Table 4.1
Vehicle ownership	low	high	Table 4.3
Cattle-dips, markets, small reservoirs	few	many	Fig. 4.2, Section 4.1.2
Agricultural extension	very poor	good	Table 4.4
Veterinary services	poor	good	Table 4.4
Other factors affecting needs satisfaction			
Sanitation facilities	very poor	good	Table 4.8
Immunization	good	very good	Section 4.3.5
Clinic service	insufficient	good	Section 4.3.5
Educational facilities	poor	good	Section 4.1.5
Use of unprotected water sources	high	low	Table 4.6

important are these resources relative to the effects of infrastructure and services?

5.3.1 *Crop production*

Despite negligible soil and rainfall differences between Zviyambe and Chiweshe, we estimate that household crop production in Zviyambe is over ten times higher than in Chiweshe in an average year. No other variables differ by such an order of magnitude between the two tenure systems, and it is undoubtedly crop production which is the primary determinant of the level of needs satisfaction in the two areas. Both retentions and sales are higher for Zviyambe households. Why is household crop production substantially higher in Zviyambe? Crop production at a household level is dependent on two variables: area planted, and yields achieved. Area planted per household is three times greater in Zviyambe, and in an average year the yields in SSCFAs in the Save catchment are 3,5 times higher than yields in CAs (see Table 5.3 for comparisons of agricultural production).

Table 5.3

DIFFERENCES BETWEEN CHIWESHE AND ZVIYAMBE IN AGRICULTURAL PRODUCTION

	Chiweshe	*Zviyambe*	*Original data*
Crop production			
Household crop production	very low	high	Section 2.2
Area planted per household	low	high	Table 2.1, Section 2.1
Yields	low	high	Section 2 2
Normal retentions	low	high	Section 2.2
Sales	low	high	Section 2.2
Draught power	little	plenty	Table 2.2, Section 2.6.1
Use of manure	limited	extensive	Section 2.6.2
Use of fertilizer	very limited	limited	Section 2.6.2
Agricultural equipment	very limited	extensive	Table 4.3
Vegetable gardens and fruit trees	few	plenty	Fig. 2.4, Section 2.2
Livestock production			
Household livestock holdings	poor	good	Tables 2.2, 2.3
Mortality rates	very high	high	Table 2.4
Calving rates	very low	low	Table 2.4
Off-take	very low	low	Table 2.4

Area planted to crops is chiefly a function of land available for planting and the inputs and labour that a household has at its disposal. There is not a shortage of land for crop production in Chiweshe, in so far as physical constraints to crop production are concerned, and the present cropping area could be expanded by about three times if exhaustion of soil fertility could be overcome. We have identified the lack of draught power as a constraint (Section 2.6), and this is probably one of the key reasons why Chiweshe farmers plant smaller areas. However, the lack of draught power is related to the poor condition of the draught animals and the limited number of animals per household. This, in turn, is related to lack of grazing. Thus, there is a land shortage, at least under the present agricultural system (i.e. with livestock being an important component of the system). Grazing land would be insufficient even if the grass resources were more carefully managed. A natural resource, land, is one of the crucial variables in determining the degree of needs satisfaction. A further reason for the limited extent of cropland per household in Chiweshe is that soil-nutrient inputs are limited and insufficient to cover large areas, especially since much of the area that could be planted has been abandoned because of nutrient depletion. Zviyambe farmers have larger herds and hence more manure for soil fertilization, and this is reflected in the present use of manure; furthermore, the manure in Zviyambe is probably of a higher quality. Zviyambe farmers also have greater access to a range of agricultural equipment (cultivators, harrows, etc.).

Although differences in yields between Chiweshe and Zviyambe may be partially related to availability of the land (insufficient cattle for manuring the fields), yield differences between the tenure systems are probably more determined by infrastructure/services components (e.g. markets, agricultural extension, availability of fertilizer) than by availability of land.

It is not only grain crop production that differs between the tenure systems: a greater number of Zviyambe households have vegetable gardens and fruit trees and this is directly reflected in the recorded food intake.

5.3.2 *Livestock production*

Some of the linkages between crop production and livestock have already been discussed, but livestock are also related to needs satisfaction in other ways. Firstly, livestock sales generate cash for various purchases, and. secondly, livestock are directly related to food needs because they are a source of protein. Livestock are able to contribute substantially to needs satisfaction in Zviyambe because household livestock holdings are sufficiently large. Furthermore, stocking rates match the carrying capacity. This is reflected in lower livestock mortality rates, higher calving rates and higher off-take. Lack of land in Chiweshe has led to severe overstocking. On the other hand, herd size per household in Chiweshe is very restricted. Once again it is ultimately the availability of land which is a crucial variable in determining needs satisfaction; however, there are other contributory factors, such as lack of sufficient veterinary control.

5.3.3 *Other infrastructure/services factors*

As indicated above, agricultural production is a key factor in determining the level of satisfaction of food and health needs. However, there are a number of infrastructural/services components which have a non-agricultural linkage to needs satisfaction (Table 5.2).

1) Sanitation

The absence of sanitation facilities has a direct effect on health. It was observed that in Chiweshe the prevalence of unsanitary conditions coincided with a high occurrence of diarrhoea.

2) Transportation

Basic needs can only be met by bringing together resources and people. There is a better transport infrastructure and greater vehicle ownership in Zviyambe compared to Chiweshe. Of the expressed needs in both areas, about 25 per cent related to the improvement of roads and transport services.

3) Health care

Although health is a basic need in itself, it can also affect the satisfaction of food needs. An absence of adequate health care can clearly put severe constraints on the amount and quality of labour available for the household farming activities, an important consideration in view of the labour-intensive nature of the farming systems. Visits to households by village health workers are extremely limited and the people-to-clinic ratio in Chiweshe ranks with the least developed countries in the world (McHale and McHale, 1978).

4) Education

Education, both formal and nonformal (village health extension, access to radios, agricultural extension, farmers' and women's groups, etc.), undoubtedly plays a role in shaping people's activities, and hence their level of needs satisfaction. Zviyambe farmers are generally better educated, both formally and nonformally. This probably plays a contributory role, not only in their ability to be more productive and in their better perception of environmental problems, but also in their attitudes and practices with regard to nutrition, sanitation, etc.

5) Water provision

Water is not piped on a large scale anywhere in the rural areas of Zimbabwe, and much reliance is placed on surface-water supplies, both for people, livestock and vegetable gardens. In the study area, the only perennial river is the Save, and even its surface flow is much reduced at the end of the dry season. Other perennial surface supplies are rare, but Zviyambe does have a large number of small dams (all of which were empty after the 1982/3 drought). Zviyambe households are much better provided with boreholes and pumps, whereas in Chiweshe most households have to rely on unprotected surface-water supplies. These differences probably account in part for the higher frequency of vegetable gardens in Zviyambe and the higher levels of diarrhoea in Chiweshe.

5.4 The effects of the production systems on the household economy

In Chiweshe only 20 per cent of households stated that agriculture was the sole source of income. Many households relied on migrant labour remittances and income from the informal sector (mostly craft work). Per capita incomes in Chiweshe are well below Z$100 per annum. Farming activities provide only marginally for families' needs. The drought is by no means the only reason for widespread poverty, for even in a good year production is low, and cash generated from agriculture is minimal (Truscott and Pambirei, 1983). In Zviyambe, where production is higher, many more households rely on agriculture for income (Table 5.4), and the higher income levels are reflected in, for example, vehicle ownership and type of housing (see Table 4.9).

Is there a lack of commitment to agriculture in Chiweshe because of the nature of the communal tenure system (i.e. a history of migrant labour, communal grazing rights, no ownership of cropping land), or is

Table 5.4

DIFFERENCES BETWEEN CHIWESHE AND ZVIYAMBE IN HOUSEHOLD ECONOMY

	Chiweshe	*Zviyambe*	*Original data*
Agriculture as a major income source	rare	general	Table 4.12
Urban-based income as major income source	general	rare	Table 4.12
Informal sector as an income source	general	rare	Table 4.12
Heads-of-households absent	many	few	Table 4.11

it poor land productivity that results in people having to rely on other cash inputs? It seems likely that the latter is the more important reason for agriculture assuming a less important role in the household economy. If, for example, one contrasts Chiweshe with the communal areas in the Honde valley, there is almost no comparison (Technosynesis, 1984). In the Honde valley, which lies in Natural Region I, agriculture based on dry-land cropping is a viable enterprise. Much of the landscape has been converted to well-planned and contoured cropping land. This landscape contrasts strongly with that of Chiweshe where there is a mix of cropping land and abandoned fields, with the latter more prominent. In the Honde valley, livestock are unimportant, and crop production dominates the household economy. Using data collected in Chamutsa Ward, in the south of Save CA, Technosynesis (1984) has compared the households having access to irrigation with the adjacent households relying on dryland farming. This area falls in Natural Region V, and therefore irrigation is essential if crop production is to be a viable enterprise. The presence or absence of irrigation determines the socio-economic conditions of Chamutsa households, including such variables as household size, family structure, degree of polygamy, income source and income per capita.

We propose, therefore, that the socio-economic differences between Zviyambe and Chiweshe are mostly a result of the differences in agricultural production. In Chiweshe, agriculture is not a profitable activity, land is in short supply, and agricultural infrastructure and services are limiting. Hence people have to rely on outside remittances and craft work to obtain even a very low degree of needs satisfaction. We have also indicated that the Chiweshe farmers have a very exploitative attitude towards the environment. This attitude is probably also related to low degree of viability of agricultural enterprises.

The poor agricultural productivity and consequently low income levels lead to an outward migration of young, active, often male, individuals. The depletion of human capital leads to increasing poverty as even the short-term absence of such a group can cause a disproportionate drop in production levels for the household. Activities which might be passed over or badly done include clearing new lands, cultivation of food or cash crops, maintenance of vegetable gardens, and craft activities. This is especially likely when it is men who migrate, leaving tasks to women in addition to their usual allocation of work (collection of firewood and water, child rearing, vegetable production and staple food crop production). In the informal discussion groups, women clearly articulated their extra workload. The situation is further complicated by the fact that until the Age of Majority Act was passed in 1982, women were minors in the eyes of the law, and since the Act was passed the status of women has not changed sufficiently in the eyes of the State or of many men. The result is that credit and extension, for example, are largely directed towards men, even though women might have been making the major agricultural decisions in their husbands' absence. In the discussion groups women stressed the need for extension workers to work with women.

The institution of migrant labour also has ambiguous consequences for the type, amount and use made of family income. Although we have indicated that remittances form a sizeable portion of per capita income, the actual amount — less than Z$15 per capita per year (Truscott and Pambirei, 1983) — could not go very far in ensuring needs satisfaction. In addition, there would appear to be much discontent on the part of women because of the irregularity of remittances (see Table 4.12) and the nature of remittances (men returning from town having spent a considerable proportion on goods which were selected with little perception of the needs of the family at home).

5.5 Environmental degradation as a consequence of the non-satisfaction of needs

One aim of this work was to assess the nature and magnitude of environmental degradation entailed by persistence of widespread poverty. To what extent does the non-satisfaction of needs lead to environmental degradation?

The tenure system which is most degraded is that in Chiweshe (Table 5.1), and this is the area where needs satisfaction is at the lowest level. Environmental degradation is in the form of deforestation, overgrazing, loss of soil fertility, erosion, siltation and changing hydrological conditions. The human activities which are undertaken in an attempt to satisfy certain aspects of basic needs are responsible for this environmental degradation. These activities are mostly related to agricultural production and include poor cultivation techniques, use of marginal cropland, and overstocking. Low crop yields, mostly as a result of loss in soil fertility, induce people to cultivate more land, which leads to decreased area for grazing and calls for more draught power. Deforestation is primarily a result of the opening up of new cropland but is also determined by the need for wood products. Most of the environmental consequences relate to decreased plant cover (woody plant cover, crop cover, and pasture cover), which is the first visual sign of desertification. The decrease in plant cover results in erosion, siltation and hydrological changes.

5.6 The implications of environmental degradation for needs satisfaction

What are the implications of environmental degradation for achieving sustained satisfaction of basic human needs in respect of food and health (Table 5.5)? Theoretically, it is possible that a human activity might be altering an environmental component to a major degree but that the component could still be capable of meeting the resource demands for the foreseeable future. A further question of interest is whether it is natural environmental limitations or man-induced environmental limitations which are most critical in the study area?

5.6.1 *The degradation trends and their consequences*

1) Overgrazing

Overgrazing has many repercussions for needs satisfaction: cattle are in poor condition and herd productivity is low; and trampling and removal of the grass resource indirectly affects needs satisfaction because of its subsequent effects (erosion, siltation, lack of thatching grass).

2) Deforestation

There is ample evidence from the study area that deforestation is directly affecting needs satisfaction. The time required for wood collection has increased considerably, a different range of species is being used, and people are having to switch to other fuel and building resources (see Du Toit *et al.* (1984) for a detailed analysis).

3) Loss of soil fertility

We do not have soil nutrient data, but many of the respondents had decreased yields and considered that this was due to soil depletion. Depleted soils usually have lower organic matter and are compacted,

Table 5.5

THE EFFECTS OF ENVIRONMENTAL DEGRADATION IN CHIWESHE ON FOOD AND HEALTH NEEDS SATISFACTION

Environmental trend	*Direct and indirect effects*
Overgrazing	Low livestock productivity Removal of grass leads to erosion
Deforestation	Lack of wood products
Loss in soil fertility	Decrease in yields Abandonment of land, new land being opened up (often marginal land) Leads to deforestation and erosion
Erosion	Decrease in productive capacity of the soil leads to siltation
Siltation	Long-term viability of present and proposed irrigation schemes
Hydrological changes	Decrease in surface water supplies Downstream erosion and flooding?

thus making them more susceptible to erosion. Loss of soil fertility indirectly results in increased erosion because of the practice of abandoning land which has been depleted of nutrients and moving to another piece of land. This practice also results in excessive deforestation because of the clearance of trees.

4) Erosion

The link between erosion and future needs satisfaction is obvious, and is indicated by the analysis in Chapter 3 which concludes that soil resources will be too degraded to meet the increased agricultural requirements of the next generation adequately. The erosion problem probably stems mostly from poor cultivation methods and the overgrazing of previously cultivated areas.

5) Siltation

The environmental degradation in the upper Save catchment gives cause for concern about the long-term viability of planned water resource developments on the Save. Two major dams are proposed for the Save, and these should supply a proposed irrigation scheme which, if developed, will be the largest irrigated settlement scheme in Africa. It is proposed that over 60 000 people will be settled on the scheme, and material benefits of the scheme will be substantial (Atkins Land and Water Management, 1983). Thus, environmental degradation in the upper Save catchment could seriously affect the viability of this scheme, and hence downstream needs satisfaction. Similarly, throughout the country many small-scale irrigation schemes can only be implemented once siltation rates are lowered. The problem of siltation is particularly important when one considers that irrigation development is probably the development which can most substantially improve the level of needs satisfaction (see Chapter 7).

6) Hydrological changes

We have demonstrated various hydrological changes that have occurred in the area, and these, at least at a local level, have an effect on needs satisfaction. For instance, perennial springs have dried up and water must be collected from greater distances. Because of lack of data we have not investigated downstream effects of the changes in the Save river flow. There is some evidence that downstream areas are now more prone to flooding, and it has been noted that bank erosion is substantial and has resulted in the destruction of roads built too close to the bank (Van Oosterhout and Campbell, 1985).

As can be seen, environmental degradation has many consequences for needs satisfaction. The degradation trend not only affects agricultural production, but also affects income gained from various informal activities — much craft work relies on natural resources which are no longer readily available, such as reeds from river banks and springs.

5.6.2 *Man-induced limitations or natural limitations?*

In the previous section we indicated that all aspects of environmental degradation have negative consequences for needs satisfaction. However, this should not be taken to mean that the present low levels of needs satisfaction result from environmental degradation. Rather, it is basic limitations, social and natural, which have set up a vicious cycle of: non-satisfaction of needs leads to environmental degradation which leads to further non-satisfaction of needs, and so on. As has been indicated, there is a land-population imbalance in Chiweshe, and this has been a major factor in initiating the degradation trend. Soil fertility is inherently low in the study area and costly fertilizer inputs (in view of the lack of sufficient livestock for manure) have to be maintained if soil nutrient levels are not to be depleted. Furthermore, the whole study area is not well suited to crop production because of the low and unreliable rainfall. In addition to these limitations of natural resources, there are, as has been indicated, many social factors which limit production and which have played a role in initiating the cycle of land degradation and non-satisfaction of basic needs.

The ultimate cause of the present situation goes back to the division of the country along racial lines, which was legalized in the Land Apportionment Act of 1930 (as described in Section 1.3.1).

5.7 Conclusions

It has been shown that the people in Chiweshe have a low degree of needs satisfaction and that this is chiefly related to the low productivity of the agricultural practices, particularly crop production. One of the chief differences between Chiweshe and Zviyambe is in population density, and this has many repercussions for agricultural production. The differences in the production systems lead to many other differences, both socio-economic and environmental. In Chiweshe, many households do not see agriculture as an important source of income. It is also in this area where deforestation, overgrazing and the abandonment of lands is most severe. It is the agricultural production system and the population–resource imbalance which has set in motion a vicious circle of land degradation, further non-satisfaction of basic needs, further land degradation, and so on. Although the agricultural production system in relation to land availability is seen as a central problem, and will be even more so in the future, it has been indicated that infrastructural and service components have also played a major role in determining the productivity of the systems in question. Apart from this indirect link via production between infrastructural and service components on the one hand and needs satisfaction on the other, these components also have various direct links (e.g. clinics, transport, boreholes).

The important effects of agricultural production on both needs satisfaction and environmental degradation suggests that, by promoting policies to improve production, we can both improve needs satisfaction and check environmental degradation. Furthermore, it has been shown that environmental quality, both natural and man-induced, has numerous consequences for needs satisfaction, and that, therefore, by ensuring policies that will promote environmental quality, the level of needs satisfaction will improve.

Irrespective of the origins of the cycle of environmental degradation and poverty in the CAs, it is clear that the farmers in these areas, as shown by our analyses, are trapped within constraints imposed by a land shortage, by many years of underinvestment on the part of authorities, and by an environment that is not well suited for dry-land crop production. The prognosis can only be gloomy—it is just a matter of time before the area will be characterized by desertification and starvation. In Chapter 7 we examine policies and programmes aimed at reducing poverty in an environmentally sustainable manner.

6 POPULATION–RESOURCE INTERACTIONS

C. A. M. Attwell

6.1 Introduction

The population of Africa south of the Sahara is growing faster than that of any other region in the world. Of the 130 or so underdeveloped nations in the world, most are now beginning to show declining rates of growth, except for those in Africa. Growth rates of sub-Saharan Africa, already the poorest region in the world, now exceed even the rest of Africa. Countries such as Nigeria, Malawi, Botswana and Zimbabwe appear likely to exceed growth rates of 3,5 per cent per annum, a dubious distinction already achieved by Kenya.

A medium variant projection of the Zimbabwe population (United Nations, 1981) gives a 71 per cent increase for the 15-year period 1985–2000, to attain a figure close to 15 million. With existing land pressures, and with such capacity for population growth, it becomes critical that any analysis of basic needs satisfaction be seen in the light of population projections against human carrying capacities.

Kay (1975) estimated that, as early as 1969, parts of Zimbabwe were some 40 per cent in excess of critical levels of safe carrying capacity, based on a relationship between population density and agricultural potential. In his analysis, the Save region itself fell into a 'grossly overpopulated' category. Indeed, most of Zimbabwe's lowveld areas are considered to fall within systems where the potential carrying capacity is exceeded by current population levels (Harrison, 1983). Harrison maintains that existing population levels could not be supported in the long term even if all cultivated land in these regions grew nothing but food crops.

It is relatively simple to arrive at an acceptable definition of carrying capacity, but notoriously difficult to arrive at valid figures for carrying capacity. Slesser and Hounam (1981) advance a definition meeting most requirements: 'the number of persons sharing a given area of territory, who can, for the foreseeable future, sustain a given physical standard of living, utilising energy, land, water, skills and organisation.' This is the sense in which the term is used here. Human carrying capacities are primarily functions of the resource base, but can be profoundly altered by changing political or economic factors. Acceptable estimates of carrying capacity levels should ideally be based on energy flow analyses (Slesser and Hounam, 1981, p. 30), but methodology for such approaches often requires data seldom available for underdeveloped countries. Other workers have based estimates on the area of land needed per capita for average subsistence, coupled with data on the extent of arable land and cultivation periods (see Brush, 1975). In this study, estimates of carrying capacity have been based on calorific requirements and arable potential. In projecting carrying capacity against crude indices of carrying capacity, a major criticism relates to the insularity of the exercise, in that little consideration is given to a more holistic prognosis, involving, for example, the development of the national economy. Such factors, whilst outside the scope of this study, should be borne in mind.*

6.2 Demography of the Condo area

6.2.1 *Methods*

Questionnaire data were used to estimate levels of fertility (Brass *et al.*, 1968); thereafter, a standard schedule of age-specific fertility was utilized for the projections (Shorter and Pasta, 1974). Child mortality was estimated via the methods of Sullivan (1972) and Brass (1976), which led to the adoption of an appropriate model life table. Projections were performed according to the cohort component method (Shorter and Pasta, 1979).

Three 'variants' with respect to total fertility rate were considered for the projections: a) a constant rate of 6,32 (the value revealed by the questionnaire); b) fertility decline to 4 by year 2018; and c) fertility

**Editor's note*: A number of group members, namely R. H. Loewenson, D. T. Pankhurst and, to a lesser extent, M. J. Swift, questioned the realism of dealing with population–resource interactions for a relatively arbitrarily defined area. Loewenson and Pankhurst went further, and suggested that solutions to the present problems must be looked at not only within the present production system. They suggested that the issue of population — and other aspects of the present report such as income — must be viewed in terms of the national and international political economy, particularly the distribution of wealth and the structures which sustain that distribution. [B. M. C.]

decline to 2 by 2018. An infant mortality rate of 60 per 1 000 was assumed; this was revealed by the questionnaire, and substantiated by United Nations (1983) projections. The latter projections were taken as a basis for changing the mortality levels in the projections.

Levels of emigration for projections ranged from that detected by the questionnaire (15,9 per cent), to levels based on projections for the urban/rural population of Zimbabwe (Tsui, 1979). Initial population levels were based on the 1982 national census figures. Projection assumptions are summarized in the footnotes to Fig. 6.1.

6.2.2 *Population structure*

Table 6.1 presents results on household size and associated dependency data. The size of the average household is an indication of social and economic conditions, and may provide a basis for comparisons between rural and urban situations (Kuznets, 1978). In the Condo, household structure resembles the pattern found in much of Africa, with large families, a high proportion of young children, very few old people, and a high female-to-male ratio between the ages of twenty to thirty-five. This biased sex ratio indicates considerable migrancy; indeed, about 25 per cent of heads-of-households were absent for over half the year. In Chiweshe, nearly 60 per cent of the labour force (defined arbitrarily as the number of residents aged ten to sixty, excluding school attenders) was under the age of fifteen.

With respect to the dependency ratios of Table 6.1, ratios are shown for 'residents and non-residents' as well as for 'residents' only, as it was considered that 'non-residents' would contribute towards the dependence of related residents, either financially or in terms of material inputs. High dependency ratios are, of course, not unique to the Save region (Whitsun Foundation, 1983a). Classically, high dependency ratios place severe demands on social services. The problem is exemplified by a medium variant projection for Zimbabwe giving a 170 per cent increase for the school-age population (5–19) between the years 1980 and 2010 (Whitsun Foundation, 1983a). Such a dependency burden imposes a high consumption-to-production ratio and a heavy work-load on the adults before children themselves become productive. This may contribute to the inability of some families to meet their children's food, education and health needs.

Figure 6.1 · POPULATION PROJECTIONS AND CARRYING CAPACITIES FOR CHIWESHE WARD

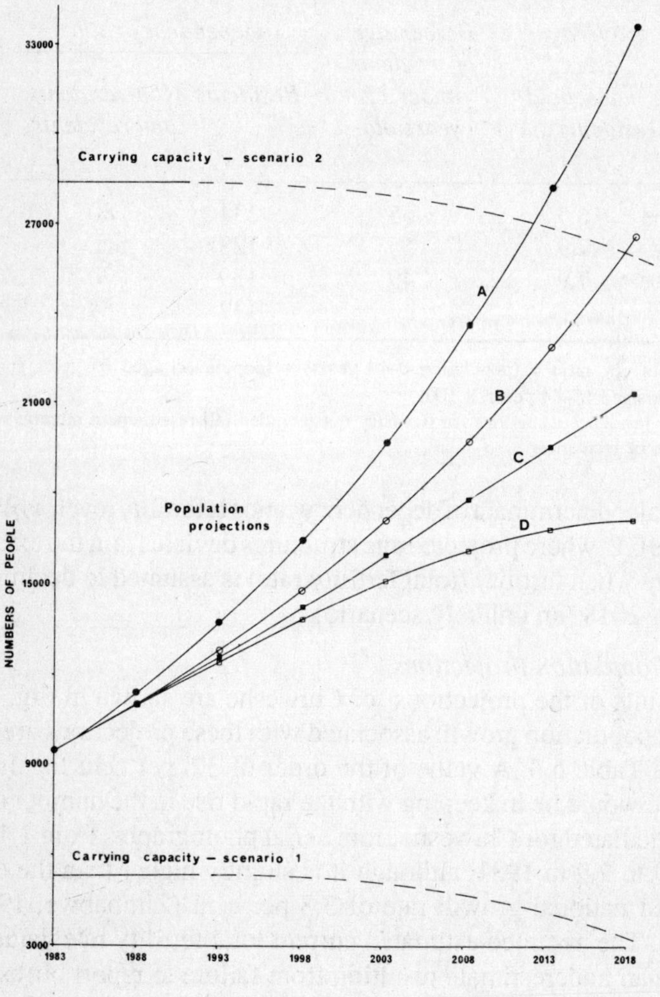

A: Total Fertility Rate = 6.32. Constant mortality, no migration.
B: Total Fertility Rate = 6.32. Decreasing mortality, rising migration.
C: Fertility declines to 4 by 2018. Decreasing mortality, rising migration.
D: Fertility declines to 2 by 2018. Decreasing mortality, rising migration.

Table 6.1

HOUSEHOLD SIZE AND DEPENDENCY DATA FOR THE CONDO AREA

Study site	*Average size of household (residents only)*	*Percentage of residents under 15 years old*	*Dependency ratio*		
			Residents	*Non-residents and residents*	*n*
Chiweshe	6,5	55	134	80	113
Romsley	7,9	55	128	n.a.	54
Zviyambe	7,9	49	139	70	58
Condo			132	84	225

Dependency ratio = (population 0–14 years) + (population aged 65 years and above) + (population aged 15–64 years) × 100.

n.a. = residence categories for Romsley not recorded (all resettlement members are legally required to be resident).

A major determinant of dependency ratios is fertility level, as illustrated in Table 6.2, where projected age structures deviate from the existing age structure when fertility (total fertility rate) is assumed to decline to 2 by the year 2018 (an unlikely scenario).

6.2.3 *Population projections*

The results of the projections for Chiweshe are shown in Fig. 6.1. The rates of population growth associated with these projections are summarized in Table 6.3. A value of the order of 3,6 per cent for the rate of increase would be in keeping with the rapid rise in the number of homesites calculated for Chiweshe from aerial photographs, from 1,1 per km^2 in 1969 to 2,9 in 1981, although it is slightly higher than the officially accepted national growth rate of 3,3 per cent (Zimbabwe, 1982b and 1983b). The assumed estimated current total fertility rate value of 6,32 may be an underestimate resulting from failure to report births, as total fertility rates for Zimbabwe have variously been estimated to lie between 6,7 and 8,0 (Whitsun Foundation, 1983a). The contribution of the relatively low total fertility rate value to the rates of increase will be partially offset by the adoption of mortality levels which may be below the national average.

Table 6.2

PROJECTED AGE STRUCTURES FOR THE CONDO AREA IN THE YEAR 2018

	Age structure as a percentage of total population	
Age class (years)	*1983*	*2018*
0–14	47	30
15–64	50	66
65+	3	4

Assuming total fertility rate declines to 2 by 2018, and based on an initial structure as sampled.

Table 6.3

PROJECTED AVERAGE RATES OF POPULATION INCREASE IN THE CONDO AREA OVER THE PERIOD 1983–2018

Projection assumption	*Rate of natural increase*	*Intercensal rate of increase*
A	3,6	3,6
B	3,6	3,0
C	3,0	2,3
D	2,4	1,7

Under differing assumptions for migration, fertility and mortality (see Fig. 6.1). The rates are averages both over time and for Chiweshe, Romsley and Zviyambe combined.

The initial population levels in the Condo correspond to densities shown in Table 6.4. These figures should be set against data available for the province and for Zimbabwe: the average population density in the rural areas of Zimbabwe is 19 persons per km^2 (Whitsun Foundation, 1983b), with Manicaland being the most densely populated province, where several CAs exceed densities of 70 people per km^2.

If one assumes the most likely projection pattern to lie somewhere between the patterns resulting from assumptions B and C (see Table 6.3), then one may compare projected densities with those set out in the Chavunduka Commission.* This Commission allowed for 16,3 people per km² in resettlement areas by 1990, and an increase from 24,8 people per km² in CAs in 1980 to 25,8 people per km² in 1990 (Whitsun Foundation, 1983b). For both Romsley and Chiweshe, the initial populations (1982 census data) already exceed the projected overall 1990 densities anticipated by the Chavunduka Commission; projections from this study for 1990 indicate approximately 28 people per km² for Romsley, and 36 people per km² for Chiweshe — densities roughly 1,7 times and 1,4 times, respectively, those anticipated by the Commission for resettlement areas and CAs generally. The 663 households planned for Romsley (Zimbabwe, 1981c) are equivalent to 3 978 people, a figure which has already been exceeded by at least three times, largely as a result of the problems of illegal squatting in this resettlement area.

Table 6.4

INITIAL POPULATION DENSITIES ASSUMED IN THE CONDO AREA COMPARED WITH AVERAGES FOR MANICALAND FOR EQUIVALENT LAND-USE SYSTEMS

	Density (people per km²)		
	Condo (1982 census)		*Average for Manicaland*
Communal area	29,0	(Chiweshe CA)	58
Resettlement area	22,5	(Romsley)	13
Small-scale commercial farming area	16,1	(Zviyambe SSCFA)	13

Manicaland data from Whitlow (1980).

6.3 Carrying capacity implications

Three scenarios were considered in arriving at different estimates for carrying capacity. In the first scenario, an estimate was based on the number of people the Condo could support in the long term under existing agricultural practices. About 60 per cent (20 000 ha) of Chiweshe has been cleared for cultivation, but much of this has lost the capacity for crop production owing to nutrient depletion and soil erosion

*Zimbabwe (1982) *Report of the Commission of Inquiry into the Agricultural Industry* [Chairman: G. L. Chavunduka] (Harare, Govt. Printer).

(see Section 3.6 and Section 3.5). Further, in a livestock-dependent system, it is imperative that a major portion of the available land be set aside as grazing area for cattle, because these animals provide, among other benefits, both manure and draught power. In 1981/2, the total cultivated area of Chiweshe Ward was only about 3 400 ha, or 10 per cent of the Ward, assigned to specific crops as indicated by Table 6.5 (Truscott and Pambirei, 1983; see Table 2.1 for comparable data for 1982/3). This Table reflects yields for the 1981/2 drought year (in fact, not as adverse as that experienced in 1982/3), and is thus adequate for carrying-capacity estimates, as it indicates productivity under poorer conditions. Despite this, the yields may yet be optimistic, as no account is taken of storage loss; such losses may be considerable (Kalff, Downing and Smith, 1985).

Dietary energy supplies can be considered the principal nutritional indicator; if dietary energy is adequate, protein will generally be adequate as well (Abercrombie and McCormack, 1976). Conversely, if there is an energy deficit, where protein supplies are in excess of requirements, part of the protein supply will be diverted to energy sources and will thus not be available for essential protein functions. Thus, using energy contents of available crops as a nutritional indicator, $4{,}88 \times 10^9$ kcals are available per year in Chiweshe (Table 6.5). Pimentel and Pimentel (1979) give 2 500 kcal as the minimum average daily energy requirement per person. This figure agrees with estimates from Save CA where the average calorific intake per person was 2 000 kcal per day, based on both the good season of 1980/1 and the poor season of 1981/2 (Technosynesis, 1984). Using a daily requirement of 2 500 kcal, and based solely on energy considerations and under existing agricultural practices, Chiweshe Ward as a whole would be able to sustain 5 352 adults in the long term, equivalent to 16 people per km^2. This density compares favourably with those proposed by Whitlow, 1980 (see Table 6.6).

The second carrying-capacity scenario assumes that the combination of crops remains unchanged (as for scenario one), but that the area of land under cultivation increases to 25 per cent of the total land. This can be regarded as a maximum for a livestock-dependent system; Technosynesis (1984) has shown that in Regions III and IV about 75 per cent of the land is required for cattle. In addition, scenario two considers the possibility that yields will be increased, equivalent to those of the large-scale

Table 6.5

CROP YIELDS AND ASSOCIATED ENERGY VALUES FOR CHIWESHE WARD AS A BASIS FOR CARRYING CAPACITY ESTIMATE (SCENARIO 1)

Crop	*Area planted (ha)*	*Percentage of total cultivated*	*Yield (kg per ha per year)*	*Total production for Chiweshe (kg per year)*	*Energy content (kcal per kg) (after Pimentel and Pimentel, 1979)*	*Gross available energy (kcal per year)*
Maize	1 530	45	500	765 000	3 550	$2{,}72 \times 10^9$
Sorghum and millet	1 020	30	500	510 000	3 300	$1{,}68 \times 10^9$
Groundnuts	510	15	200	102 000	3 900	$3{,}98 \times 10^8$
Other food crops	100	3	250	25 000	3 500	$8{,}75 \times 10^7$
Cash crops	238	7	n.a.	n.a.	n.a.	n.a.
Total	3 398	100				$4{,}88 \times 10^9$

Data for 1981/2; source: Truscott and Pambirei (1983).

Table 6.6

ESTIMATED SAFE CARRYING CAPACITIES (PEOPLE PER KM²) FOR REGIONS III AND IV

Agro-ecological region	*Whitlow (1980)*	*Whitsun Foundation (1983b)*
III	15	45
IV	10	30

Assuming current agricultural practices.

commercial farming sector. Ivy (quoted in Whitsun Foundation, 1983b) has provided frequency distributions for maize yields (tonnes per hectare) from different natural regions, over a five-year period. The data were drawn from Intensive Conservation Areas in large-scale commercial farming areas over a period of generally good seasons (1970/1–1974/5), and can thus be taken as approaching maximal figures attainable. From these distributions, it was estimated that Region III was capable, on average, of 2,7 tonnes maize per hectare compared to 2,2 tonnes per hectare for Region IV. If one takes a mean value of 2,45 tonnes per hectare as a basis for an estimate of carrying capacity for scenario two, one must consider the effects of poor seasons on this yield, as a carrying-capacity estimate must be based on the maximum productivity under the poorest conditions. Truscott and Pambirei (1983) report a figure of 1 140 kg per ha from Chiweshe after the good rains of 1980/1. In the following poor season, this yield fell to less than 500 kg per ha, a reduction of 56 per cent. Further, for the season of the severe drought of 1982/3 an average yield of less than 250 kg per ha was recorded in the Condo, with more than 50 per cent of households experiencing total crop failure (see Section 2.2). If one, nevertheless, disregards the 1982/3 season as being too extreme, and uses the data of Truscott and Pambirei, then the value of 2,45 tonnes per hectare must be reduced by a factor of 56 per cent to give 1,08 tonnes per hectare of maize. Thus, to arrive at increased yields consistent with those of the commercial sector, the maize yield shown in Table 6.5 must be doubled. If the other crop yields are increased by the same factor, then $2,53 \times 10^{10}$ kcal per year will be available (Table 6.7), being the requirements for 28 055 people. This is equivalent to 86 people per km².

Table 6.7

YIELDS AND AVAILABLE ENERGY FOR CARRYING CAPACITY ESTIMATE OF CHIWESHE WARD (SCENARIO 2)

Crop	*Area planted (ha)*	*Yield (kg per ha per year)*	*Total production (kg per ha)*	*Gross available energy (kcal per year)*
Maize	3 672	1 080	3 965 760	$1{,}41 \times 10^{10}$
Sorghum and millet	2 448	1 080	2 643 840	$8{,}72 \times 10^{9}$
Groundnuts	1 224	432	528 768	$2{,}06 \times 10^{9}$
Other food crops	240	540	129 600	$4{,}54 \times 10^{8}$
Total				$2{,}53 \times 10^{10}$

Yields have been increased by a factor of 2,16 (see Section 6.3) and area under cultivation by a factor of 2,40. The present area under cultivation, 3 400 ha, is assumed to increase to 8 175 ha, being 25 per cent of the total area of Chiweshe.

Scenarios one and two are indicated against the population projections of Fig. 6.1. Clearly, under existing agricultural practices (scenario one), population levels are grossly in excess of the carrying capacity, and this difference will become more marked as population increases. Under scenario one, population will remain in excess of carrying capacity irrespective of any fertility variant considered. This imbalance will further reduce the carrying capacity: as ecosystem degradation proceeds, so, too, must the carrying capacity decrease. Herd size increases with population (whether as cause or effect), and the effects of overstocking are well documented. In addition, the soil resource in most regions of the Save will be too degraded to meet the increased agricultural requirements of the next generation (see Sections 3.5 and Section 3.6). This anticipated reduction in carrying capacity is arbitrarily indicated by the downward trends of carrying capacity scenarios in Figure 6.1.

However, it is clear that the dietary deficit must in part be compensated for, if one is to explain the persistence of current levels of population. Food deficiencies are met largely through remittances. In Chiweshe, one third of the household income is derived from employment outside the Ward (Truscott and Pambirei, 1983; see also

Table 4.11); this fraction may rise to two-thirds in drier regions of the Save (Technosynesis, 1984). Further, as most residents are children (Table 6.1), the gross nutritional requirement of the population will be proportionately decreased. Despite these considerations, over 20 per cent of Chiweshe children are moderately or severely undernourished according to weight-for-age criteria (Table 4.14).

In scenario two, carrying capacity would be attained by about the year 2010 under existing patterns of fertility; this assumes that the agricultural innovations of scenario two must not only initially bring the carrying capacity into line with current population levels but thereafter 'keep abreast' of population growth.

In scenario three, the same yields as for scenario two are assumed, but the land under cultivation is now considered to increase to 50 per cent of the total area available. This is a most unrealistic scenario, as not only would it mean a livestock-independent system (thus requiring a technological revolution), but it would place severe strain on ecological systems, given the land capability classes in the area (see Section 1.5.4). Although this would double the carrying capacity of scenario two, it cannot be considered as a serious alternative. Regardless of crops planted and arable extent, farmers in Regions III and IV must be constrained by unreliable rainfall and the prospect of frequent crop failure, and hence must rely on livestock production to some extent (see agro-ecological recommendations in Section 1.5.4). It would thus appear that human nutritional needs in Chiweshe could be met in the long term only under scenario two, and then only provided that fertility reduction followed a pattern similar to that of projection D (see Table 6.3).

It is accepted that the above projections can only serve as crude indices of 'limits to growth', as little consideration has been given to future industrial development outside the Condo area. Were such development to provide a 'buffer' to the system, it would again require concomitant fertility reduction for any long-term ameliorative effect.

Population projections against carrying capacity for Romsley suggested a similar prognosis to that outlined above. The situation for Zviyambe (again based on population projections) appears less pessimistic, provided again that agricultural innovation is accompanied by fertility reduction. However, Zviyambe cannot be viewed as an ecological or sociological entity, and it will undoubtedly receive considerable socio-ecological feedback from processes at work in surrounding areas. Degradation of peripheral areas would, in all likelihood, lead to increased pressures on Zviyambe (as is already occurring to some extent).

In conclusion, the carrying capacity of Chiweshe will be exceeded by a factor of over three by the turn of the century, under existing agricultural practices and current fertility patterns. A similar situation can be assumed for most of the Condo region. Even under extremely favourable agricultural developments (scenario two), carrying capacity will be exceeded early in the next century, unless there is a pronounced reduction in human fertility. One can anticipate increasing use of marginal land; Froude (quoted in Whitsun Foundation, 1983b) notes that an arable land ratio of approximately 4 ha per cultivator is a traditional ratio, and that when numbers of cultivators rise, the tendency is not for holdings to decrease proportionately, but for the otherwise landless to cultivate in essentially non-arable areas. Utilization of marginal land for arable purposes has proved disastrous for low-rainfall regions of Africa, not the least being the destruction of grazing land (Sinclair and Fryxell, 1985).

The people of Chiweshe generally felt that although human population pressure in Chiweshe is becoming intolerable the problem can be readily overcome through resettlement. This chapter suggests that, under present population dynamics, equitable distribution of better-quality land (for example, through resettlement schemes) contributes but little to the alleviation of both human poverty and environmental degradation, as any redress in carrying capacity so attained will be rapidly 'tracked' by a fast-growing population, until the original population–resource imbalance prevails (see also Section 7.2).

6.4 Attitudes to contraception

Clearly, the transition to reduced fertility is a complex process, of which adequate contraceptive facilities is but one facet. This study did not address itself in any way to the determinants of high fertility in the Condo, and the inclusion of data on attitudes to contraception, does not necessarily imply that fertility is determined solely by levels of contraceptive awareness and availability. However, as fertility reduction is critical to any desirable population–resource scenario in the Condo, it is pertinent to consider the people's perceptions on contraceptive issues.

Many women in the Condo saw an interval of two to three years between births as desirable on both health and nutritional grounds, but, nevertheless, wished to maximize the number of children within this constraint of birth interval. The justifications were typical of those described in underdeveloped countries, and related to insurance against child mortality and to investment in children as a source of labour and old-age security: 'To have many children is to be rich' (Zviyambe respondent). There was a marked difference in attitude between women of Chiweshe and Zviyambe towards what was considered a desirable birth interval (Table 6.8), with Zviyambe women citing a longer birth

Table 6.8

WOMEN'S ATTITUDES TOWARDS BIRTH INTERVAL AND FAMILY SIZE, TOGETHER WITH CLINIC VISITS DURING THE LAST PREGNANCY

Attitudes towards a 'good' birth interval
(0,001 < p < 0,01; d.f. = 2)

	Chiweshe (n = 108)	*Zviyambe (n = 53)*
1–2 years	63	42
2,5–3,5 years	26	54
4+ years	11	4

Attitudes towards the 'best' number of children to have
(n.s.; d.f. = 3)

	Chiweshe (n = 110)	*Zviyambe (n = 57)*
4–5	20	21
6–8	34	40
9+	15	18
'many'	31	21

Women visiting clinic during last pregnancy
(0,001 < p < 0,01; d.f. = 1)

	Chiweshe (n = 102)	*Zviyambe (n = 57)*
	59	81

Percentage of women.
Results of chi-squared analyses are shown.

interval. However, with respect to fertility goals, no difference in attitude was detected between women of Chiweshe and Zviyambe: most women wanted six to eight, or 'many', children. These views may reflect different infant mortality rates in the two areas. Ante-natal health care, as reflected by clinic visits, is much better in Zviyambe (Table 6.8).

Most methods of contraception were viewed either as anathema or with caution. This was marked in the case of steroidal contraceptives (both oral and injectable), which were rumoured to have far-reaching side-effects, even to include disease and deformity in children. In particular, men were totally opposed to contraception, apparently because this placed control of fertility largely with women: 'A man without children has no power' (Chiweshe respondent). Consequently, the minority of women who wish to take oral contraceptives must generally do so without their husbands' knowledge.

This situation points to a need for family-planning extension work to dispel contraceptive rumours which lack any medical basis, and for measures to improve the status of woman. Young couples appear to be under considerable pressure from the extended family to continue producing children; many women respondents mentioned that they are subject to the threat of divorce if their fertility is not maintained.

7 POLICIES AND PROGRAMMES

B. M. CAMPBELL AND J. A. CODD (COMPILERS)

7.1 Introduction

In this chapter we attempt to specify for Chiweshe the economic, social and environmental measures required to promote protection and improvement of the environment in harmony with the satisfaction of basic needs. From the foregoing chapters, it is obvious that a major problem to be addressed is the present and future population–resource imbalance. Interrelated social and agricultural problems to be addressed are: low crop production; low livestock production; shortage of wood products; and poor infrastructure and services. In proposing policies, a further problem emerges: restrictions imposed by traditional land rights.

Policies and programmes to combat each of the above problems are discussed; we do not discuss the measures required to combat the environmental problems separately. These environmental problems, which all revolve around reduced cover of plants and depleted soil nutrients, will be addressed largely by the policies aimed at solving the social and agricultural problems.

7.2 Population–resource imbalance

Fundamental to almost every environmental problem in Chiweshe is population pressure. The previous chapters (5 and 6) have indicated that the availability of land is one of the crucial variables determining present and future levels of needs satisfaction. For short-term planning (with, for example, five-year projection periods) it would probably be sufficient to implement policies which are aimed at improving the production systems (improved extension, accessibility to markets, etc.), but for longer-term plans (with, for example, 15-year projection periods) it would be essential to address the land–population imbalance. Our analyses have indicated that the agricultural production system must be improved to near-maximum efficiency and that 25 per cent of the landscape must be under crop production in 20–30 years' time if the projected numbers of people are to be supported by the land (see Section 6.3). There would be very little area remaining for cattle production even if some land was used for fodder production, and cattle could only play a subsidiary role in the agricultural system. However, an agricultural system based solely on crop production is not recommended because of the frequency of droughts in the area.

The land question was one of the major driving forces behind the civil war, and the major post-war rural development programme concerns resettlement. Technosynesis (1984) regarded the land–population imbalance as the main factor responsible for resource depletion and the non-satisfaction of basic needs. In their rural development proposal for Chiweshe, they were unable to see a means of accommodating the future population, and their 10-year projections for the plan are based on a static farmer population — i.e. any further increase in population must be accommodated through off-farm activities (resettlement, local industries, etc.). This is a basic flaw in an otherwise sound plan.

A number of the policies advocated in the following sections would alleviate the resource–population imbalance. Thus, for example, policies aimed at increasing agricultural productivity should directly increase carrying capacity but should also introduce local agricultural surpluses, and this will, in turn, activate various local economic activities (crafts, services, etc). The proposed public works scheme (see Section 7.5) would involve the local population, particularly during the slack periods in the farming season. It is important to realize that, whatever dry-land crop-production systems are followed, there are going to be drought years when the population will be unable to feed itself and food imports will be necessary. In these years the public works scheme should have the flexibility to expand into a 'food-for-work' scheme, and be able to absorb a major section of the population. The 'food-for-work' scheme should only be an emergency measure, providing immediate food when required and some longer-term benefits associated with infrastructural improvements.

The above types of schemes are only temporary or ameliorative; in order to tackle the population–resource imbalance in a realistic and objective manner we propose that the government give full attention to a National Population Policy. We must immediately clarify our position by stating that population policies in isolation will only have limited success when poverty remains widespread. Population policies are not a substitute for social and economic development policies. On the other hand, development programmes do not have the desired effect when populations continue to grow rapidly. Population and development policies are mutually supportive (United Nations Environment Programme, 1984). Any National Population Policy must:

(a) set proposed carrying capacities in various regions under different scenarios;

(b) set the targets for a rate of population growth which will not outstrip the nation's financial and natural resource development capability;

(c) devise methods to increase awareness, at the community level, of benefits, both socially and individually, of family planning;

(d) devise methods to increase awareness of the nature and prospect of environmental degradation, its relationships to population pressures, its likely impacts on proposed development projects and on people's lives, and how local action can improve the situation.

Our second proposal is for the development of intensive agricultural development on the basis of irrigation. Water is the major limiting resource in the study area. At a local level, irrigation development is the single most important means of ensuring food security and of increasing the land's carrying capacity. This is indicated by the major development proposal for Chisumbanje where it is projected that the farmer population of an area is to rise from 108 in year one to 13 000 in year nineteen (Atkins Land and Water Management, 1983). It is not, of course, possible to propose a scheme of such magnitude for Chiweshe, as water supplies are limited, and even the Chisumbanje scheme is threatened by the degradation trends in the upper Save catchment. Technosynesis (1984) proposes a small-scale scheme (70 ha) on the southern boundary of Chiweshe and 50 ha of irrigated area around wells. We give full support to such proposals.

There are various other developments which could alleviate the excessive farmer populations in the CAs. These are briefly examined:

(a) Industrial development. No industrial development capable of creating significant migration flows towards the urban areas or rural growth points is foreseeable in the future, though there would appear to be some opportunity for small-scale rural industrial development (Zimbabwe, 1983b, p. 5).

(b) Improved wages/pensions. Raising wages and improving family security, and thereby creating a committed proletariat, could result in the weakening of ties to rural areas of the 'holiday farmer' and a greater commitment to farming by those who remain in the rural areas. However, no significant changes in wages/pensions are foreseeable given the vast surplus of unskilled labour and a relatively stagnant economy.

(c) Resettlement policies. The present resettlement programme is making no progress towards relieving land pressure and is not even keeping pace with present population increase (Kinsey, 1983; Whitlow, 1984). Our analysis of Romsley Resettlement Scheme indicates the population problems (see Sections 6.2 and 6.3).

7.3 Low agricultural production

Many measures need to be introduced if agricultural production is to be maximized, and if soil loss is to be minimized. In this drought-prone environment it is dangerous to rely solely on crop production; hence our proposals envisage a mixed farming system. However, there is insufficient grazing land to support the draught and manure inputs required for productive cropping; hence our scenario is based on outside inputs into the system.

The need to control the grazing and movement of cattle, and to intensify crop production on the inherently fertile red soils, implies a radical reorganization of present land use. Radical reorganization of present land use would not only affect agricultural production: because of the consolidation of homestead sites, services such as water supplies, schools, clinics and market facilities would be easier to organize. A limited capital outlay on such services should be considered in any land-use programme, and should form part of a public works programme (see Section 7.5). In general, reorganization of land use would involve identifying arable blocks, areas for grazing schemes, woodlot sites, water points, village sites, etc., and planning the proposed land use to ensure optimal utilization of the resources. Enough is now known about Chiweshe to make this a very simple task. Community participation in such an important planning exercise would be necessary from the outset (see Section 7.6).

Only 10 per cent of the land has red soils. About 50 per cent of the land has been cleared for cultivation, but less than 15 per cent is utilized for cultivation in any one year. Cropping land should be organized into continuous blocks and to a large extent should be concentrated on the red soils. An arable holding of five hectares per household, as proposed by the resettlement plans for Natural Regions III and IV, will give a return of about Z$400 per capita per year to the household (see Zimbabwe, 1981c). Of the five hectares, two would be fallow. If these figures are used, about 50 per cent of the people presently in Chiweshe could be accommodated on the red soils. The other 50 per cent would remain on the white soils. In total, about 20 per cent of the land would be under cultivation, leaving about 75 per cent of the land for grazing schemes.

The land for grazing could, if rehabilitated and improved, have a carrying capacity of 8 hectares per livestock unit, which would mean that the total livestock units in the Chiweshe study area could be in the order of 1 500, i.e. 2,3 livestock units per household (at present densities of people). This figure, or at least the predicted one livestock unit per household at the turn of the century, is too low for livestock to make a significant contribution to soil replenishment and draught power (Sandford, 1982). There are two possibilities: Certain villagers may opt for mechanical support and thereby lose their right to own cattle. For example, the farmers on the red soils could be given mechanical support and subsidized fertilizer input. These soils are the more difficult to plough with oxen, and they do not require as much fertilizer input, whereas the white soils need organic inputs to maintain fertility, and are easier to work. If such an arrangement could be organized, the farmers on the white soils would have about five livestock units per household (at present densities of people). The other possibility is the management of communally-owned livestock, a certain percentage of which would be available to the owners for ploughing. The above broad analysis of hectareage available and consequent production per household indicates the major steps that need to be taken if the problems are to be solved. It is all too simple to state merely that production must be increased — the land most suitable for intensification is limited, and whatever scenario is envisaged, there is insufficient grazing land for large livestock holdings per household.

A wealth of information has been collected during many years of research on methods of crop production and livestock management that might improve productivity and conserve soil in the CAs (see Du Toit, 1985, for a discussion of some aspects). The point to be appreciated is that technical solutions to many of the problems are available. Of prime importance is the implementation of grazing schemes, and the use of mechanical soil-conservation measures to reinforce the biological conservation which will result from improved crop productivity. Mechanical soil-conservation methods require labour and draught power to implement, and Technosynesis (1984) therefore proposes a 'mechanical unit' equipped with the necessary machinery to implement soil conservation measures. We feel that the responsibility for conservation should rest with individual farmers. Nevertheless, we do see the need for a separate unit responsible for major reclamation works. Such mechanical conservation is best undertaken within a public works scheme, which will at the same time be providing a local employment source (see Section 7.5).

With respect to increasing productivity, we have noted that livestock alone can never be in sufficient numbers to provide for the soil nutrient needs. On the other hand, the widespread use of chemical fertilizers is not economically feasible and does have environmental repercussions. A top priority should be given to seeking other ways of sustaining soil fertility. This means exploring other means of tillage, and of integrating trees into the agricultural system. The sandy soils, in particular, need input of organic matter. The opportunity to use and recycle agricultural residues more effectively is enormous and only limited by lack of incentives and appropriate research and development. The use of residues by CA farmers would involve no costs, apart from the indirect cost of utilizing the residues for soil fertilization rather than as, at present, livestock fodder. The gains involved would be a) reduced reliance on costly artificial fertilizers, and b) improved maintenance of soil fertility, which would, in turn, lead to higher crop yields and less abandonment of land (i.e. more land available for grazing).

7.4 Shortage of wood products

The shortage of wood products is now pronounced in Chiweshe and a reafforestation programme is essential. It would appear that although community or 'demonstration' woodlots are essential, the long-term aim must be to provide wood at a household level. The cultivation of woodlots could provide a permanent barrier against erosion caused either by rainfall or by wind, at the same time providing fuel and poles and a reserve of fodder for times of drought. There appears to be an urgent need for research on possible species for planting in the drier areas of Zimbabwe, both as woodlot species and agro-forestry combinations.

Apart from the cultivation of trees, it is necessary to manage the natural woodland to ensure sustained use of woodland products, and to manage the natural regeneration on abandoned fields. There has been almost no research on either of these management aspects. The natural woodlands in Chiweshe cover nearly 15 per cent of the landscape and are restricted to hills which are not even ideally suitable for livestock production. Over 50 per cent of the area consists of vegetation in various stages of regeneration, and this vegetation could undoubtedly be manipulated to improve plant cover and resource yields. Many wild fruit trees remain in the otherwise disturbed vegetation and in the planted fields. Development proposals should recognize the importance of wild fruit trees to rural households (Campbell, 1987).

7.5 Poor infrastructure and services

Public works programmes need to be designed and implemented in the CAs. The programmes should both generate employment and, simultaneously, improve the infrastructure and enhance the environment. The principal aims of such a programme in Chiweshe should be: to improve the transport network; to improve mechanical soil-conservation measures; to improve water supply (household and agricultural); to plant and manage woodlots; to erect fencing; and to build or provide building bricks for schools, clinics, cattle-dips, latrines, etc.

As has been mentioned, the public works programme must have the flexibility to change to a 'food-for-work' programme during times of drought. Public works or 'food-for-work' programmes can be expected to have profound effects. Technosynesis (1984) estimates that for a cost of approximately Z$5,5 million (though this includes a perhaps excessively high percentage of heavy equipment — Z$1 million) the following can be achieved in a four-year programme:

(a) income generation: Z$1 million allocated to wages of personnel;

(b) employment: 25 supervisory staff, each for four years, and 480 labourers each for one year (or, better, many more labourers on shorter-term contracts); and

(c) production levels:

i) area reclaimed with contour ridges — 12 000 ha;

ii) number of wells sunk — 250;

iii) feeder roads upgraded — 100 km;

iv) small dams built — 3;

v) area fenced for grazing schemes — 58 000 ha;

vi) small-scale irrigation schemes — 70 ha;

vii) improvement of natural vegetation and woodlots — 10 000 ha;

viii) protection of eroded areas and river banks — 1 000 ha;

ix) infrastructure and buildings for eight villages (200 households, including the following: upgrading/building for five schools; upgrading one clinic; construction of five community centres; construction of five market facilities).

Such a programme would have many beneficial repercussions for the residents in Chiweshe; not only would it directly raise per-capita income but it would also provide a much-needed infrastructure which would increase crop and livestock production. In addition, it would provide direct measures to ensure environmental protection. In another village in Save CA, a 'food-for-work' scheme demonstrated the viability of developing resources on a community basis. This village now boasts a major tree resource: 2 000 eucalyptus for fuel and poles, and 1 000 fruit trees.

Apart from the infrastructure provided by the above programme, it is essential that better marketing facilities be developed in Chiweshe. Local markets should serve catchment areas of up to about a twelve kilometre radius and act as load consolidation points. Access to agricultural inputs would be improved by their being available at a local market. The standing of a local market could be improved by the presence of a parastatal marketing organization.

A major problem in introducing many of the policies and programmes is that of achieving sufficient contact between the agricultural extension staff and the Chiweshe farmers; at present there is a ratio of only one extension worker to every 850 farmers in the CAs, and in the Chiweshe survey 60 per cent of the heads-of-households said that they had never been visited by extension workers. The present extension service is clearly inadequate and needs to be improved, although this improvement may be partially brought about by the extension workers disseminating new farming ideas through representatives elected by the farming communities; these representatives would receive some basic training in agricultural skills and could then relay advice to farmer's groups which would be formed to hold regular meetings and to co-ordinate the local farming operations (see Truscott and Pambirei, 1983; Bates, 1980; Plowes, 1980).

The major health problems encountered in the study area are, predictably, related to poverty and inadequate access to primary health care. The severe shortage of safe water in Chiweshe must be remedied, and the health service needs to be upgraded both in terms of clinics and village health workers. One of the aims of the public works programme should be the provision of pit-latrines.

7.6 Conclusions and general strategies

We have proposed eleven general areas in which policies and programmes must be developed in order to raise agricultural productivity, improve needs satisfaction and halt environmental degradation. Table 7.1 shows the eleven areas and the problems which they aim to overcome. Of the eleven, only one is of a national character, the National Population Policy. The other ten need to be implemented in the context of the local agro-ecological situation.

A major problem with many of the proposals outlined here is the question of communal land rights. The traditional allocation of land rights can no longer ensure the proper use of scarce resources. Various authorities (e.g. Stubbs, 1977) suggested that as land becomes in short supply it is necessary to register those landholders who were full-time farmers and to allow them more access to land resources at the expense of those people who were more securely established elsewhere but who still exercised the right to retain land in their rural areas. Mukundu (1976) identified such 'holiday farmers' as being a major constraint to production in the CAs. The new resettlement programme requires that settlers do not earn money from off-farm activities, and is the beginning of a new

Table 7.1

INTERACTIONS BETWEEN SOCIAL, AGRICULTURAL AND ENVIRONMENTAL PROBLEMS ON THE ONE HAND AND POLICIES AND PROGRAMMES ON THE OTHER

Social and agricultural problems	low crop productivity	low livestock productivity	shortage of wood products	poor infrastructure and services
Associated environmental problems	use of marginal land			
	low plant cover and erosion	overgrazing	deforestation	
	soil nutrient depletion			
Policies and Programmes				
National Population Policy	•	•	•	•
Irrigation development	•			
Re-organization of land-use	•	•		
Grazing schemes		•		
Increase agricultural productivity	•	•		
Reafforestation programme			•	
Management of indigenous woodlands		•	•	
Extension programme	•	•	•	
Improve marketing	•	•		
Improve health service				•
Public works programme (contours, dams, woodlots, dips, boreholes, fencing, etc.)	•	•	•	•

land ethic among the rural poor. Technosynesis (1984) suggests that the flow of investments to development projects should be linked to local communities' framing new rules for access to land rights by their members. What has to be promoted in the implementation phase is the exchangeability of land rights (arable, grazing, wood-harvesting, etc.); here, every member of the community is given a share of the land resources owned in common which the member is allowed to exchange or lease. This will push enterprising farmers to improve their skills so as to be able to purchase or hire more land rights than those allocated on the basis of their membership of the community. A basic premise of this approach is that conservation is best achieved by allowing a community direct participation in the management of the resource, and by ensuring that the benefits accrue directly to the community concerned. For the implementation of such ideas, a 'land company' would be formed to manage the resource. It would be formed by members of the community and by government members actively involved in the project. Such ideas have been implemented on an experimental basis in other parts of the country, but are mostly based on the wildlife resource (Taylor, 1982).

Effective land-use planning is often inhibited by too rigid a division of responsibility among government bodies, for example, in Zimbabwe among the ministries or parastatals concerned with agriculture, water development, forestry, and transport. We would support the proposals of the Scientific Council of Zimbabwe for a land-use planning agency to co-ordinate such planning.

Crucial to any development proposal is that it is taken over by the community and that implementation is, at least partly, community-based.

> No matter how sound the public policies for environmentally responsible and socially just development, it is crucial to mobilise popular participation in their implementation and their continued evolution. Since the notion 'quality of life' has an abiding subjective aspect, it is essential that the population at large participates actively in the formulation and implementation of development programmes and projects (Tolba, 1982, p. 13).

Our survey and that of Truscott and Pambirei (1983) indicate that the people understand the issues and problems, are anxious to see improvements in their conditions, and are prepared for radical transformation — but only if they are equal participants in the process.

We have indicated in Chapter 4 that a high level of community organization already exists. These organizations can form the basis for community-directed projects. The value of community groups for directing extension advice has already been mentioned. The community organization should be the basis through which demonstration woodlots, demonstration plots and communal grazing schemes are set up.

One final aspect of general strategy concerns the role of women. The Chiweshe area has a high degree of male migrancy and the heads-of-households are often women. We have indicated in our analysis the importance women now have in the rural economy, and therefore it is necessary that the advocated policies and programmes take into account the role of women. Priority attention needs to be given to establishing a programme of basic education for women. Improvement in the social status of women can play a pivotal role in accelerating the progress of societies towards their desired population, environment and development goals (United Nations Environment Programme, 1984).

REFERENCES

ABERCROMBIE, K. AND McCORMACK, A. 1976 'Population growth and food supplies in different time perspectives', *Population Development Review*, II, 479–8.

ARRIGHI, G. 1970 'Labour supplies in Southern Rhodesia', *Journal of Development Studies*, VI, 28–35.

ATKINS LAND AND WATER MANAGEMENT 1983 'Feasibility Study for the Development of 40 000 Hectares of Irrigation at Chisumbanje, Zimbabwe' (Harare, Agricultural and Rural Development Authority).

BANGEZHANO, C. K. 1983 'People–Vegetation Interactions in the Upper Sabi Catchment Area' (Harare, Univ. of Zimbabwe, Dept. of Biological Sciences, B.Sc.(Hons.) diss.).

BATES, D. B. 1980 'Important innovations introduced by agricultural extension to the tribal farmer over the last twenty years', *Rhodesia Science News*, XIV, 187–90.

BEHNKE, R. H. 1985 'Measuring the benefits of subsistence versus commercial livestock production in Africa', *Agricultural Systems*, XVI, 109–35.

BERG, A. 1981 *Malnourished People: A Policy Review* (Washington, World Bank).

BRASS, W. 1976 'Indirect methods of estimating mortality illustrated by application to Middle East and Northern African data', *Population Bulletin*, X, 125–57.

BRASS, W. *et al.* 1968 *The Demography of Tropical Africa* (Princeton, Princeton Univ. Press).

BRATTON, M. 1984 'Draft Power, Draft Exchange and Farmer Organizations' (Harare, Univ. of Zimbabwe, Dept. of Land Management Working Paper 9/84).

BRUSH, S. B. 1975 'The concept of carrying capacity for systems of shifting cultivation', *American Anthropologist*, LXXVII, 799–811.

Burki, S. J. 1980 'Sectoral priorities for meeting basic needs', *Finance and Development*, XVII, 18–22.

Callear, D. 1984 'Land and food in the Wedza Communal Area', *Zimbabwe Agricultural Journal*, LXXXI, 163–8.

Campbell, B. M. 1987 'The use of wild fruits in Zimbabwe', *Economic Botany*, XLI, 375–85.

Campbell, B. M. and Du Toit, R. F. 1988 'Relationships between wood resources and use of species for fuel and construction in the communal lands of Zimbabwe', *Monographs in Systematic Botany of Missouri Botanic Gardens*, XXV, 331–41.

Campbell, B. M. and Du Toit, R. F. (in press) 'Vegetation patterns in a semi-arid savanna area in Zimbabwe, with special reference to the influence of peasant agricultural systems', *Kirkia*.

Chavunduka, G. L. 1982 'Social aspects of conservation', *Zimbabwe Science News*, XVI, 289–92.

Chikanza, T. C., Paxton, D. and Loewenson, R. 1981 'The health status of farmworker communities in Zimbabwe', *Central African Journal of Medicine*, XXVII, 88–91.

Cleghorn, W. B. 1966 'Report on the conditions of grazing in Tribal Trust Lands', *Rhodesia Agricultural Journal*, LXIII, 57–67.

Cliffe, L. 1981 'Zimbabwe's political inheritance', in C. Stoneman (ed.), *Zimbabwe's Inheritance* (London, Macmillan), 8–35.

Dasmann, R. F., Milton, J. P. and Freeman, P. H. 1974 *Ecological Principles for Economic Development* (London, Wiley).

Du Toit, R. F., Campbell, B. M., Haney, R. and Dore, D. 1984 'Wood Usage and Tree Planting in Zimbabwe's Communal Lands' (Harare, unpubl. report for the Forestry Commission of Zimbabwe and the World Bank).

Du Toit, R. F., van Sickle, J. and Campbell, B. M. 1985 'Hydrological changes in the upper Sabi catchment', in R. F. Du Toit, 'Soil loss, hydrological changes, and conservation attitudes, in the Sabi Catchment of Zimbabwe', *Environmental Conservation*, XII, 160–1.

Du Toit, R. F. 1985 'Soil loss, hydrological changes, and conservation attitudes, in the Sabi Catchment of Zimbabwe', *Environmental Conservation*, XII, 157–66.

ELWELL, H. A. AND STOCKING, M. A. 1982 'Developing a simple yet practical method of soil-loss estimation', *Tropical Agriculture*, LIX, 43–8.

ELWELL, H. A. AND STOCKING, M. A. 1984 'Estimating soil life-span for conservation planning', *Tropical Agriculture*, LXI, 148–50.

GAPARE, R. 1976 'Problems in crop production as seen by a Tribal Trust Land farmer', *Rhodesia Science News*, X, 249–51.

GERMAN AGENCY FOR TECHNICAL CO-OPERATION 1985 'Co-ordinated Agricultural and Rural Development, Gutu District, Masvingo Province, Main Report' (Harare, Ministry of Lands, Resettlement and Rural Development).

GHAI, D. M., GODFREY, M. AND LISK, F. 1980 *Planning for Basic Needs in Kenya* (Geneva, International Labour Office).

GOULD, W. T. S. 1978 'Guidelines for School Location Planning' (Washington, World Bank, Staff Working Paper 308).

GRANT, P. M. 1976 'Peasant farming on infertile sands', *Zimbabwe Science News*, X, 252–4.

GRANT, P. M. 1981 'The fertilization of sandy soils in peasant agriculture', *Zimbabwe Agricultural Journal*, LXXVIII, 169–75.

HARDIN, G. 1968 'The tragedy of the commons', *Science*, CLXII, 1243–8.

HARRISON, P. 1983 'Land and people, the growing pressure', *People* [*Earthwatch*], X, 1–8.

HUDSON, N. W., DAVIES, F. G. AND PILDICH, A. G. 1964 'Final Report of the Sabi Catchment Investigation Team' (Salisbury, Federal Dept. of Conservation and Extension).

HUNTLEY, B. J. AND WALKER, B. H. 1983 *Ecology of Tropical Savannas* (Berlin, Springer-Verlag).

INTERNATIONAL LABOUR OFFICE 1976 *Declaration of Principles and Programme of Action for a Basic Needs Strategy of Development* (Geneva, ILO).

INTERNATIONAL UNION FOR CONSERVATION OF NATURE AND NATURAL RESOURCES 1980 *World Conservation Strategy* (Gland, IUCN).

IVY, P. 1981 *A Guide to Soil Coding and Land Capability Classification for Land-Use Planners* (Harare, Govt. Printer).

KALFF, J., DOWNING, J. A AND SMITH, T. T. 1985 'Rainfall, agriculture, livestock and human density in the dry regions of Kenya', *Journal of Arid Environments*, IX, 173–83.

KAY, G. 1975 'Population pressure and development prospects in Rhodesia', *Rhodesia Science News,* IX, 7–13.

KEECH, M. A. 1969 'Determination of trend using air-photo analysis', *Rhodesia Agricultural Journal*, LXVI, 113–22.

KENYA. CENTRAL BUREAU OF STATISTICS 1983 *Third Rural Child Nutrition Survey, 1982* (Nairobi, Ministry of Finance and Planning).

KINSEY, B. H. 1983 'Emerging policy issues in Zimbabwe's land resettlement programmes', *Development Policy Review*, I, 163–9.

KUZNETS, S. 1978 'Size and age structure of family households: Exploratory comparisons', *Population Development Review*, IV, 187–224.

MCHALE, J. AND MCHALE, M. C. 1978 *Basic Human Needs: A Framework for Action* (New Brunswick, Transaction Books).

MEIKLE, G. 1982 *Soil Erosion of Sabi Catchment and Suggested Remedies* (Harare, Ministry of Agriculture, Dept. of Agricultural, Technical and Extension Services).

MORLEY, D. AND HAUMAN, P. 1979 'The Third World: What the child eats and how this has changed our approach to malnutrition', *Nursing Times*, X, 28–35.

MUIR, K. 1981 'Crop Production Statistics 1940–1979' (Harare, Univ. of Zimbabwe, Dept. of Land Management, Working Paper 4/81).

MUKUNDU, M. Z. 1976 'Factors limiting increased crop production in Tribal Trust Lands as an extension worker sees them', *Rhodesia Science News*, X, 248–9.

NDLELA, D. B. 1981 *Dualism in the Rhodesian Colonial Economy* (Lund, Univ. of Lund, Dept. of Economics, Economic Studies XXII).

NIE, N. H. *et al.* 1979 *SPSS: Statistical Package for the Social Sciences* [computer program] (London, McGraw-Hill).

NORTON, A. 1983 'Tillage research in Zimbabwe', *Zimbabwe Science News*, XVI, 6–10.

PHIMISTER, I. 1977 'Peasant agriculture and underdevelopment in Southern Rhodesia, 1890–1914', in R. Palmer and N. Parsons (eds.), *The Roots of Rural Poverty in Central and Southern Africa* (London, Heinemann), 255–67.

PIMENTEL, D. AND PIMENTEL, M. 1979 *Food, Energy and Society* (London, E. Arnold).

PLOWES, D. C. H. 1980 'The impact of agricultural extension in the eastern tribal areas of Zimbabwe', *Zimbabwe Science News*, XIV, 197–200.

REGIONAL WATER AUTHORITY 1986 *Twenty-first Annual Report and Accounts* (Harare, Regional Water Authority).

RHODESIA 1979 *Integrated Plan for Rural Development* (Salisbury, Ministry of Finance).

RIDDELL, R. C. 1978 *The Land Problem in Rhodesia* (Gwelo, Mambo Press).

RODER, M. G. W. AND HOPLEY, J. D. H. 1973 'Investigation into systems of farming suitable for Tribal Trust Land', *Rhodesia Agricultural Journal*, LXX, 1–18.

RODER, W. 1964 'The division of land resources in Southern Rhodesia', *Annals of the Association of American Geographers*, LVI, 41–52.

SANDBACH, F. 1980 *Environment, Ideology and Policy* (Oxford, Blackwell).

SANDFORD, G. 1982 'Livestock in the Communal Areas of Zimbabwe' (London, Overseas Development Institute [Report produced for the Ministry of Lands, Resettlement and Rural Development, Zimbabwe]).

SCUDDER, T. 1973 'The human ecology of big projects: River basin development and resettlement', *Annual Review of Anthropology*, II, 45–55.

SCUDDER, T. 1980 'River-basin development and local initiative in African savanna environments', in D. R. Harris (ed.), *Human*

Ecology in Savanna Environments (London, Academic Press), 385–405.

Shorter, F. C. and Pasta, D. 1974 *Computational Methods for Population Projections with Particular Reference to Development Planning* (New York, Population Council).

Shumba, E. M. 1984a 'Reduced tillage in the Communal Areas', *Zimbabwe Agricultural Journal*, LXXXI, 235–9.

Shumba, E. M. 1984b 'Yields of maize in the semi-arid regions of Zimbabwe', *Zimbabwe Agricultural Journal*, LXXXI, 91–4.

Sinclair, A. R. E. and Fryxell, J. M. 1985 'The Sahel of Africa: Ecology of a disaster', *Canadian Journal of Zoology*, LXIII, 987–94.

Slesser, M. and Hounam, I. 1981 'A Methodology to Evaluate the Influence of Population Change, Socio-cultural Factors and Availability of Land and Physical Resources in Extending or Curtailing the Carrying Capacity of States' (Glasgow, Univ. of Strathclyde, Energy Studies Unit).

Stubbs, A. T. 1977 'The Tribal Trust Lands in transition: Land-use', *Rhodesia Science News*, XI, 181–4.

Sullivan, J. M. 1972 'Models for the estimation of the probability of dying between birth and exact ages of early childhood', *Population Studies*, XXVI, 79–97.

Taylor, R. D. 1982 'Buffer zones: Resolving conflicts between humans and wildlife interests in the Sebungwe, Zimbabwe', *Zimbabwe Agricultural Journal*, LXXIX, 179–84.

Technosynesis 1984 'Rural Development in Buhera District' (Harare, Technosynesis [Report to the Agricultural and Rural Development Authority]).

Thompson, J. G. and Purves, W. D. 1978 *A Guide to the Soils of Rhodesia* (Salisbury, Govt. Printer, Rhodesia Agricultural Journal Technical Handbook 3).

Tolba, M. K. 1977 'Foreword', in McHale and McHale, *Basic Human Needs* (q.v.), 1–2.

Torrance, J. D. 1983 'Why the drought?', *Zimbabwe Science News*, XVII, 154–6.

TRUSCOTT, K. 1985 *Socio-economic Factors Related to Food Production and Consumption: A Case Study of Households in Wedza Communal Land, Zimbabwe* (Harare, Ministry of Agriculture, Dept. of Agricultural, Technical and Extension Services).

TRUSCOTT, K. AND PAMBIREI, N. C. 1983 'Buhera–Chiweshe Ward: Report of a Socio-economic Survey 1982–83' (Harare, Ministry of Agriculture, Dept. of Agricultural, Technical and Extension Services).

TSUI, A. O. 1979 *Illustrative Functional Projections 1975–2000: Rhodesia* (Chicago, Univ. of Chicago, Community and Family Study Center).

UNESCO 1972 *Impact of Human Activities and Land-Use Practices on Grazing Lands: Savanna, Grassland, Tundra* (Paris, Unesco, Man and the Biosphere Report Series VI).

UNESCO 1975 *Impact of Human Activities and Land-Use Practices on Grazing Lands: Savanna, Grassland* (Paris, Unesco, Man and the Biosphere Report Series XXV).

UNICEF 1984 'Report on the Botswana Workshop on Clinic-based Nutritional Surveillance Systems and Integrated Data Bases' (Nairobi, Unicef).

UNITED NATIONS 1980 *Zimbabwe, Towards a New Order* (New York, UN).

UNITED NATIONS 1981 *Demographic Indicators of Countries: Estimates and Projections as Assessed in 1980, Medium Variant* (New York, UN).

UNITED NATIONS 1983 *Population Bulletin of the United Nations* (New York, UN).

UNITED NATIONS ENVIRONMENT PROGRAMME 1982 'Basic Needs in the Arab Region' (Nairobi, UNEP, Reports and Proceedings Series V).

UNITED NATIONS ENVIRONMENT PROGRAMME 1984 *The State of the Environment 1985* (Nairobi, UNEP).

VAN OOSTERHOUT, S. A. M. AND CAMPBELL, B. M. 1985 'Vegetation and Land-Use in the Communal Areas of the Middle Sabi Alluvial Plain' (Harare, Technosynesis [Report to the Agricultural and Rural Development Authority]).

VINCENT, V. AND THOMAS, R. G. 1960 *An Agricultural Survey of Southern Rhodesia: Part I: Agro-ecological Survey* (Salisbury, Govt. Printer).

WERGER, M. J. A. 1978 'Tropical grasslands, savannas, woodlands: Natural and man-made', in W. Holzner, M. J. A. Werger and I. Ikusima (eds.), *Man's Impact on Vegetation* (The Hague, Junk), 107–37.

WHITLOW, J. R. 1980 'Environmental constraints and population pressures in the tribal areas of Zimbabwe', *Zimbabwe Agricultural Journal*, LXXVII, 173–81.

WHITLOW, J. R. 1984 *Environment, Land Utilization and Development Prospects in the Peasant Farming Areas of Zimbabwe* (Keele, Univ. of Keele, Dept. of Geography, Occasional Paper 8).

WHITSUN FOUNDATION 1980 *Rural Service Centres Development Study* (Salisbury, Whitsun Foundation).

WHITSUN FOUNDATION 1983a *Population and Development Problems in Zimbabwe: A National Population Policy Study* (Harare, Whitsun Foundation).

WHITSUN FOUNDATION 1983b *Land Reform in Zimbabwe* (Harare, Whitsun Foundation).

WILLIAMS, O. 1981 'Irrigation farming in the south-east lowveld of Zimbabwe: Retrospect and prospect', *Geography*, LXVI, 228–32.

WORLD HEALTH ORGANIZATION 1978 *A Growth Chart for International Use in Maternal and Child Health Care* (Geneva, WHO).

WORLD HEALTH ORGANIZATION AND UNICEF 1978 *Alma-Ata International Conference on Primary Health Care* (Geneva, WHO).

YUDELMAN, M. 1964 *Africans on the Land* (Cambridge MA, Harvard Univ. Press).

ZIMBABWE 1981a *Zimbabwe Conference on Reconstruction and Development: Conference Documentation* (Salisbury, Ministry of Economic Planning and Development).

ZIMBABWE 1981b *Growth with Equity: An Economic Policy Statement* (Salisbury, Ministry of Economic Planning and Development).

ZIMBABWE 1981c 'Romsley Intensive Resettlement Area: Final Report' (Harare, Ministry of Agriculture, Dept. of Agricultural, Technical and Extension Services, Planning Branch).

ZIMBABWE 1982a *Social Conditions in District Council Areas in Zimbabwe* (Harare, Central Statistical Office).

ZIMBABWE 1982b *Transitional National Development Plan 1982/83–1984/85 Volume 1* (Harare, Ministry of Economic Planning and Development).

ZIMBABWE 1983a 'Wedza Baseline Study: Summary and Results' (Harare, Ministry of Agriculture, Dept. of Agricultural, Technical and Extension Services, Socio-economic Unit).

ZIMBABWE 1983b *Transitional National Development Plan 1982/83–1984/85 Volume 2* (Harare, Ministry of Economic Planning and Development).